G.2

D0305586

7

Deformation of Solids

Applications of Mathematics Series

Editor: Alan Jeffrey
Professor of Engineering Mathematics
University of Newcastle-upon-Tyne

WILLIAM F. AMES
Numerical Methods for Partial Differential Equations

S. BARNETT and C. STOREY
Matrix Methods in Stability Theory

F. J. BAYLEY
Heat Transfer

T. J. M. BOYD and J. SANDERSON
Plasma Dynamics

D. J. EVANS
Numerical Methods

C. D. GREEN
Integral Equations and Methods

I. H. HALL
Deformation of Solids

JEREMY HIRSCHHORN
Dynamics of Machinery

ALAN JEFFREY
Mathematics for Engineers and Scientists

S. KELSEY
Matrix Mechanics of Structures

B. PORTER
Synthesis of Dynamical Systems

WILLIAM K. ROOTS
Control Systems Engineering

H. RUND
Variational Methods

H. N. V. TEMPERLEY
Statistical Thermodynamics

Deformation of Solids

I. H. HALL

Lecturer in Physics
University of Manchester
Institute of Science and Technology

NELSON

THOMAS NELSON AND SONS LTD

36 Park Street London W1
P.O. Box 336 Apapa Lagos
P.O. Box 25012 Nairobi
P.O. Box 21149 Dar es Salaam
P.O. Box 2187 Accra
77 Coffee Street San Fernando Trinidad

THOMAS NELSON (AUSTRALIA) LTD
597 Little Collins Street Melbourne 3000

THOMAS NELSON AND SONS (SOUTH AFRICA) (PROPRIETARY) LTD
51 Commissioner Street Johannesburg

THOMAS NELSON AND SONS (CANADA) LTD
81 Curlew Drive Don Mills Ontario

THOMAS NELSON AND SONS
Copewood and Davis Streets Camden N.J. 08103

First published in Great Britain 1968

© I. H. Hall 1968

Illustrations in text by David Middleton

17 177066 8 (Paper)

17 176067 0 (Boards)

Made and Printed in The United Kingdom by
The Universities Press Limited, Northern Ireland

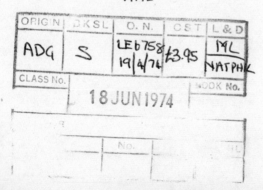

Preface

Several years' experience of research into the mechanical properties of materials, both in industry and in research associations, has convinced me that the traditional 'Properties of Matter' course for undergraduates provides an inadequate training for the person who is later to specialize in this subject. The traditional presentation omits recent developments such as the theory of rubber elasticity; it does not emphasize that the mathematical theory of elasticity is only applicable in cases where certain conditions are satisfied, or that assumptions are made about the experimental behaviour of materials in developing this theory; and, worst of all, it cannot be developed to include advanced topics such as three-dimensional states of stress and strain, finite strain, or anisotropic materials.

To overcome these difficulties we need to develop an entirely new approach, which this book attempts to do. The mathematical treatment is kept as simple as possible, which frequently involves proving a theorem rigorously for two dimensions, and assuming an analogous result for three. At the same time, however, the student is encouraged to think in three dimensions. Wherever possible geometrical concepts, which can be visualized, are used in preference to abstract mathematics. Throughout, the assumptions which are made about the response of real materials to deforming stresses are emphasized, and the restrictions which must be introduced to simplify the mathematical development of the theory are explained. The notation is such that more advanced treatments, involving matrix and tensor methods, will follow naturally.

The first part of the book deals with methods of specifying stress and strain and leads to the assumptions about the behaviour of materials which must be made if these quantities are to be simply related. The second part is concerned with the experimental behaviour of real materials and their deviation from the assumptions made in the first part, and gives methods of describing this deviation. A final chapter qualitatively relates the observed behaviour to the molecular structure.

In my view, the contents constitute a minimum syllabus in the subject which all physics students should cover. It is to be hoped that they will cover more, and that, in particular, the subject matter of the last chapter will be considerably developed. For this reason, this chapter has been deliberately restricted to a qualitative discussion of physical ideas, emphasizing the differences in the origin of elastic forces in different types of material. Numerous specialist text-books are available which develop these ideas quantitatively.

The book was written with the needs of physics students particularly in mind, but should also be of value to engineering and materials science students. It should be suitable for a student either nearing the end of his first undergraduate year, or beginning his second. While he is working through it, other parts of his course will be increasing his knowledge and ability in mathematics and physics. For this reason the mathematical treatment becomes more abbreviated as the book progresses, and physical concepts which he might be expected to have encountered elsewhere by the time he reaches the later parts of the book are introduced without discussion.

I would like to acknowledge my gratitude to Professor L. R. G. Treloar, who kindled my interest in this subject; to Professor H. Lipson for allowing me to develop my ideas in undergraduate courses at the University of Manchester Institute of Science and Technology; and to Professor M. M. Woolfson, but for whose encouragement I might not have written this book. I am indebted to my colleagues Dr. A. Kaye and Miss A. Sutherland, both of whom have read the manuscript and made many useful suggestions; to Mr. M. H. Lewin of the publishers for his editorial assistance and co-operation, and also to my wife for invaluable assistance in correcting the manuscript and proofs.

I. H. HALL

Contents

1

Introduction

When forces act on a body they alter its size and shape. For example, in Fig. 1-1 a cylindrical rod is clamped vertically at one end and an axial

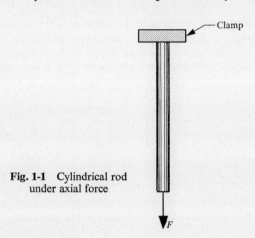

Fig. 1-1 Cylindrical rod under axial force

force is applied to the other. This force will change the length of the rod, and possibly also the radius. In this book we are concerned with the relationship between changes in the dimensions of a body and the forces acting on it, and how this relationship depends on the material from which the body is made.

This is a subject of interest and importance in many scientific disciplines. Engineers must design structures, such as bridges, which will not deform dangerously under the loads they encounter in service. On the other hand, the technologist concerned with the shaping of materials needs to deform bodies permanently from one shape to another and must know the forces developed in his processing machinery in doing this. Physicists and materials scientists are interested in the relationship between the molecular structure of a material and its mechanical properties. They want to know how differences in the properties of two materials depend on their molecular arrangements, and how to alter the properties of a material to make it more suitable for a particular purpose.

This book has been written primarily for students of physics and materials science. However, before the relationship between the mechanical properties of a material and its molecular structure can be discussed, we must know how to define and measure these properties, and how they differ from one class of material to another. Accordingly, most of the book will be devoted to these topics. Their relation to the molecular structure of the material will be considered only in the final chapter, and there only in a general and qualitative manner.

In this chapter we outline the argument used; in subsequent chapters we shall discuss the various stages in detail.

If one parameter can be found which defines the force acting on a body and another can be found which defines the deformation, their relationship can be illustrated by a graph. For the rod shown in Fig. 1-1, the axial force can be represented by a single number and the deformation by the change in length of the rod. Graphs are plotted of axial force against change in length in Fig. 1-2 for a rubber rod of 0·5 cm² cross section and 5 cm long, and in

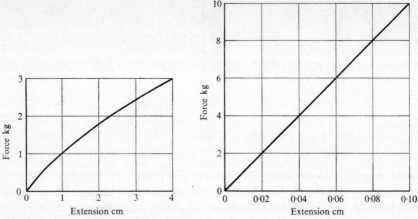

Fig. 1-2 Load–extension curve—rubber rod 5 cm long, and 0·5 cm² cross-sectional area

Fig. 1-3 Load–extension curve—steel rod 100 cm long, and 0·005 cm² cross-sectional area

Fig. 1-3 for a steel rod of 0·005 cm² cross section and 100 cm long. The magnitudes of forces and deformations are so different for the two materials that they cannot be shown on the same graph, but how far are these differences due to the different dimensions of the rods, and how far are they due to the different properties of the materials from which the rods are made?

For this particular experiment the effect of dimensions can be determined quite easily. Suppose equal axial forces are applied to two identical rods, each of unextended length l. They will both extend by the same amount, δl, as illustrated in Fig. 1-4(a). If we join the two rods end to end and again apply an axial force F, each component rod will experience the same axial force as before, and so each will again extend by δl [Fig. 1-4(b)]. The parameter used to represent deformation should therefore have the same value as before, but the total extension of the combined rod will be $2\delta l$. However, if the extension $2\delta l$ is divided by the unextended length $2l$, the resultant

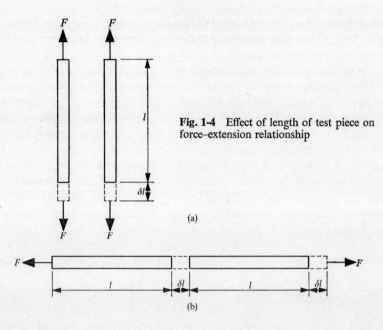

Fig. 1-4 Effect of length of test piece on force–extension relationship

quantity will have the same value, $\delta l/l$, as when the rods were stretched individually. This quantity is therefore independent of test piece length and is referred to as the *strain*. A more general definition of this term is given in Chapter 2.

The effect of cross-sectional area can be deduced in a similar manner. Suppose equal axial forces were applied to two identical rods of rectangular cross section, and distributed uniformly over the end faces, as in Fig. 1-5(a).

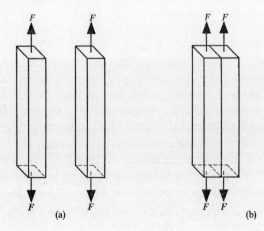

Fig. 1-5 Effect of area of cross section of test piece on force–extension relationship

Now suppose the two rods are joined along one of their side faces, as in Fig. 1-5(b). Each will experience the same axial force as before and so the parameter used to represent force should be unaltered. The total force applied to their end faces will be doubled, but so will the area of cross section; so if the force is divided by the area of cross section, the resultant quantity will have the same value as when the rods were stretched individually. This quantity is therefore independent of the area of cross section of the test piece and is referred to as the *stress*. Again, a more general definition is given in Chapter 3.

In Figs. 1-6 and 1-7, stress is plotted against strain† for the rubber and steel rods. The magnitudes of the quantities still differ greatly between the two materials, but it is now certain that the difference between their mechanical properties is genuine, and not a spurious effect arising from different test piece dimensions.

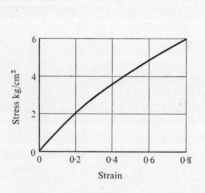

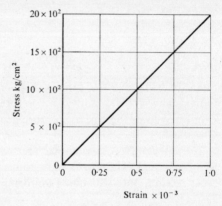

Fig. 1-6 Stress–strain curve—rubber rod   **Fig. 1-7** Stress–strain curve—steel rod

However, this is only one of many ways of deforming a material and plotting a graph of the relationship between force and deformation. For example, we could twist one end of a rod relative to the other and plot a graph of the twisting couple against the angle of twist. Alternatively, we could clamp a bar horizontally at one end, apply a vertical force to the other, and plot a graph of the force against the deflection of the free end. (For reasons which will become apparent in a later chapter, the effects of the test piece dimensions cannot be eliminated in these cases as easily as they were above.) Another way of deforming a body is to immerse it in fluid under pressure, when the change in volume per unit volume is plotted against the

† The definition which has been used to calculate the strains in these figures is strictly only applicable to strains much smaller than those in Fig. 1-6. However, this is not important in the present argument, and will be discussed in more detail in Chapter 2.

fluid pressure. Many other examples could be cited, all leading to different relationships between the forces acting on the body and the deformation they produce.

Are all these relationships independent of each other? If they are not, can we determine a few, more basic, relationships and use these to calculate the deformation resulting from applying any system of forces? If we can find such relationships, we shall have a method of comparing properties of different materials which is not only independent of the dimensions of the test piece, but also of the particular type of experiment that has been performed.

To answer these questions we need more general definitions of stress and strain than those already given. The definition of strain is suitable only for a deformation in which the length of the rod is changing; a different definition would be necessary for other types of deformation. However, consider a small cubic element of material within the rod. In general, when the rod is deformed the edges of this cube will change in length, and the angles between the edges will alter, as in Fig. 1-8. So far, our definition of strain

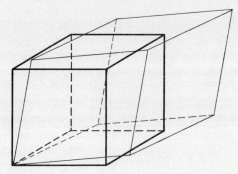

Fig. 1-8 Distortion of small cubic element of deformed body

considers only the change in length of the edge parallel to the axis of the rod, and ignores the changes in other dimensions. If we can develop it to include the changes in the lengths of all the edges and the angles between them, then it will apply to any type of deformation. In Chapter 2 we develop such a definition and explore the mathematical properties of the quantity defined, but these mathematical properties are simple only if the changes in length and angle of the elemental cube are small, i.e., the ensuing theory will be restricted to small deformation.

So far we have only defined stress in terms of axial force acting along a rod. We can now generalize this definition as follows. If the rod in Fig. 1-9 is cut through at the plane AA', the two halves will fly apart. Thus the molecules lying on one side of this plane must exert a force on those on the other side. It is this internal force which prevents the rod separating into parts before it is cut.

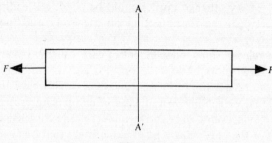

Fig. 1-9

Now consider a small cube of material within the rod, as in Fig. 1-10. Internal forces due to the surrounding molecules will act on its faces, as shown. From the previous definition, the stress on the cube is the magnitude of the internal force acting on the face normal to the rod axis, divided by the area of this face. Had external forces been applied differently from those shown in Fig. 1-9, then internal forces would act on the other faces of the cube in Fig. 1-10 and would not necessarily be normal to these faces. We must therefore extend our definition of stress to include the forces on all faces, and their directions of action. We develop such a definition in Chapter 3, and demonstrate that the mathematical properties of this quantity are the same as those of strain.

When stress and strain are defined in this way, they require several numbers to specify them completely. For example, the deformed cube of Fig. 1-8 will require six quantities to specify its size and shape: the lengths of the edges intersecting at any corner and the three angles between these edges. In general, both stress and strain require six numbers to specify them completely. However, we see in Chapters 2 and 3 that any state of stress or strain can be resolved into three components, each of which can be specified by a single number.

The strain produced by each of these stress components will be determined in Chapter 4. Each stress component produces a strain which can also be represented by a single number, provided that these strains do not depend on the direction in the material in which the stress component is applied (such materials are called *isotropically deformable*). Thus one graph can relate stress and strain for each one of these stress components. Furthermore,

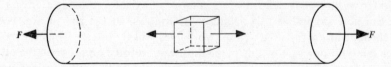

Fig. 1-10   Internal forces in rod with axial tension

for two of the stress components the graphs of stress against strain are identical.

Thus, we can analyse any stress into three components and determine the strains produced by these components, knowing only two stress–strain relationships. Provided attention is restricted to small strains in elastically isotropic bodies, we need only know these two basic relationships to determine the deformation in any body resulting from the application of any system of forces. These are the relationships which should be studied experimentally.

Unfortunately, they concern changes in the dimensions of an elemental cube of material, and the internal forces which act on its faces, neither of which quantities are directly measurable. For experimental purposes, these relationships must be expressed in terms of deformations occurring and forces applied in experimentally realizable situations, such as the examples already cited. Using certain assumptions, we can calculate the changes in the dimensions of a body caused by systems of forces, as is done in Chapter 5. We can use experimental investigation of these changes to check the validity of our assumptions. The results of such experimental investigations are discussed in Chapter 6.

What assumptions must we make? First, that the stress is proportional to the strain; and second, that the strain depends only upon the value of the stress and not upon any other quantity (such as the length of time for which it has been applied). Our first assumption allows us to express each relationship by a single number, the constant of proportionality. The two numbers so determined are called the *elastic moduli* and define the elastic properties of the material.

Since the method of analysing elastic deformations is only valid at small strains in elastically isotropic materials, we must restrict experimental investigation to these conditions. The experiments then show that the first assumption—that stress is proportional to strain—is valid. The second assumption, however, cannot be confirmed experimentally. If the strain is measured at different intervals of time after imposing a stress, a variation with time will be detected, and therefore it is not uniquely defined by the stress. This behaviour is called *visco-elasticity*. Experimental investigations such as these also reveal that the restriction of the theoretical treatment to small strains and elastically isotropic materials is completely unrealistic. Many materials can be deformed beyond the limitations of the small strain theory, many are anisotropic.

Materials can now be classified according to their mechanical properties. For example, one of the elastic moduli of rubber is smaller than that for metals by a factor of about 10^4. Furthermore, rubber will recover its original dimensions after very large deformations, whereas very much smaller deformations in metals are irrecoverable. From the theoretical analysis these are genuine differences between the material properties, and are not spurious effects resulting from the choice of test piece dimensions, or the type of experiment performed. We are therefore justified in explaining these differences in terms of the molecular structure of the materials.

The molecular structures of certain classes of material are known, and the way in which these structures determine mechanical properties can be described qualitatively, and, in some cases, quantitatively. However, detailed quantitative theories are beyond the scope of this book, and for these the specialist texts dealing with particular classes of material should be consulted.

2

Specification of Strain

2-1 State of strain

When a system of forces acts on a body, it will produce two effects: an acceleration (unless the forces are in equilibrium), and changes in shape and size. The changes in shape and size are called the state of strain. How is this specified?

In the previous chapter, we defined strain as the change in length divided by the unextended length. However, we noted that such a definition is restricted to deformation in one dimension of the body. A more general definition and specification will now be developed to include all three dimensions, but for simplicity we will start by considering two-dimensional sheets of material.

2-2 Strain and rigid body movement

In mathematics, we commonly locate a point in space by specifying its perpendicular distance from three mutually perpendicular planes. These distances are called the cartesian coordinates of the point. In a two-dimensional sheet of material, two coordinates suffice. One might think that any changes in these coordinates would specify the deformation of the sheet. However, Fig. 2-1 shows that a point† P, in a sheet of material ABCD, can be displaced to P′, either by moving the whole body rigidly to A′B′C′D′, as in Fig. 2-1(a), or by deforming the body to A′B′C′D′, as in Fig. 2-1(b). The former case is known as *rigid body movement*. Thus two coordinates are unsatisfactory because they will not distinguish between deformation and rigid body movement.

Suppose that, instead of a point, we have a line PQ in the sheet. If the sheet is moved rigidly [as in Fig. 2-1(c)], the line stays the same length; but if the sheet deforms [Fig. 2-1(d)], the line changes in length. Such changes in length indicate deformation and not rigid body movement.

However, we have still not completely solved the problem. Rigid rotation

---

† In the diagrams in this book, points are marked to represent particular particles in the material. As the particle moves, so the point moves. Similarly, lines denote particular lines of particles, and changes in shape or length of the line record changes in the relative positions of particles. Diagrams of bodies before and after deformation are often superimposed and, in this case, particles and lines of particles are denoted respectively by unprimed symbols (e.g., P) and heavy lines before deformation, and by primed symbols (P′) and medium-weight lines after deformation.

of the sheet [Fig. 2-1(e)] will cause the line to rotate; so also will extension in a direction inclined to that of the line [Fig. 2-1(f)]. To distinguish between these causes, we must draw a second line inclined to the first. If the angle between the lines is unchanged [as in Fig. 2-1(g)], then rigid body rotation has occurred. Changes in angle [Fig. 2-1(h)] can only occur as a result

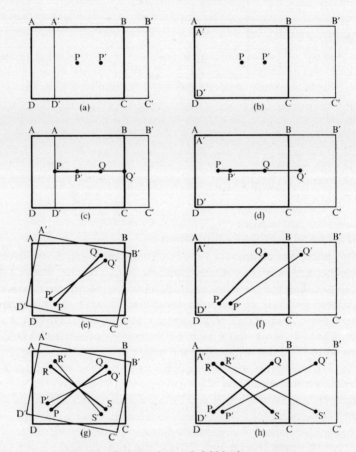

Fig. 2-1 Deformation and rigid body movement

of deformation. Thus, to measure the deformation of a two-dimensional sheet of material, and to distinguish between this and rigid body movement, we need to draw at least two inclined lines in the sheet and measure the change in angle between them and their changes in length.

2-3   Shear and extensional strain

Customarily, we draw the two lines specifying the deformation of a sheet of material so that they are mutually perpendicular in the undeformed state.

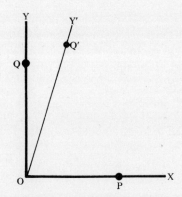

Fig. 2-2   Shear and extensional strain of OY direction

Two such lines are shown as OX, OY in Fig. 2-2. Consider two particles P and Q, one on each line. Now suppose that, after deformation, the sheet is moved rigidly so that the particle at O returns to occupy the same position in space that it occupied before deformation.† The sheet is also rotated rigidly so that the line of particles defining OX is in the same direction in space before and after deformation. Further, suppose the particles at P and O are the same distance apart before and after deformation. Then (from Fig. 2-2) the line of particles defining OY has rotated to OY′, and has extended so that the particle Q now occupies the position Q′.

It is difficult to define the strain in a body in terms of the displacements of its constituent particles in a way to satisfy all conditions. Considerable simplification is achieved by considering only small strain, which is the only state that we shall discuss in detail. A deformed metallic body will recover its original shape and size on releasing the deforming forces (*elastic* behaviour) only if the strains are very small; at larger strains, the deformation would be permanent (*plastic* behaviour). The theory we shall develop will therefore be adequate for the elastic deformation of metals, but will not be valid for the large elastic deformations which are possible in rubber, or for plastic deformation.

Two quantities are necessary to specify the displacement of the particle Q (Fig. 2-2): (a) the extension of the line OQ, and (b) its rotation. From the first of these quantities is defined

$$\textit{extensional strain} = \frac{OQ' - OQ}{OQ} \qquad (2\text{-}1a)$$

From the second is defined

$$\textit{angle of shear} = \text{angle } QOQ' \qquad (2\text{-}1b)$$

† Unless specifically stated otherwise, this rigid body motion will be applied in all cases considered in this chapter.

It must be clearly understood that these definitions apply only to small strain. The general definitions are more complicated, and these represent their limiting cases as the strain approaches zero. From these definitions and Fig. 2-2, we see that, even though the strain is small, the displacement of Q can be large if OQ is large.

An assumption not previously mentioned has been used in drawing Fig. 2-2. The line of particles which defines OY, which was straight before deformation, is assumed to be straight afterwards. This is not necessarily so. We have made another assumption in the definition of extensional strain. In Eqn. (2-1a), no restriction is placed on the position of Q, implying that the extensional strain is independent of this position. Again, this is not necessarily so. We will consider the implications of these two assumptions later.

The example of deformation discussed above is a very simple one. Generally, the deformation will extend OP as well as OQ. Also, rigid rotation of the deformed sheet, to make the direction of OX the same as before deformation, is not usual. The situation, taking account of these two points, is then as shown in Fig. 2-3. Again, we assume that straight lines remain straight

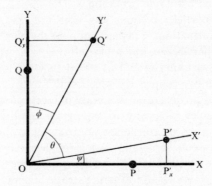

Fig. 2-3   Shear and extensional strain of OX and OY directions

after deformation, and that the extensional strains are independent of the positions of P and Q.

Three quantities are necessary to specify the relative displacements of P and Q:

(a) the extensional strain in the direction OY $= \dfrac{OQ' - OQ}{OQ}$ (2-2a)

(b) the extensional strain in the direction OX $= \dfrac{OP' - OP}{OP}$ (2-2b)

(c) the angle of shear in the XY plane (this is the difference between the initial and final values of the angle XOY) =
$\frac{1}{2}\pi - \theta$ (2-2c)

We use the phrase 'relative displacements' to indicate that, although these three quantities define the shape and size of the deformed sheet completely, they do not give its position and orientation in space. We have already used the convention that a rigid body movement is applied so that the particle O occupies the same position in space as before deformation. Also, either the angle ψ or ϕ must be known to determine the orientation of the sheet.

If we are to apply the techniques of cartesian coordinate geometry, we need to express these strains in terms of displacements along the coordinate axes. However, we have not yet defined coordinate axes, although two lines of particles, mutually perpendicular before deformation, have been labelled OX and OY. On deformation, these lines rotate to OX' and OY', and are no longer mutually perpendicular. Hence, *lines of particles are not suitable for use as axes.* The difficulty is overcome by defining the *directions in space of the lines before deformation* as the coordinate axes.

Since the strains are small, their components along the coordinate axes can be obtained quite easily. Perpendiculars to OX and OY are drawn from P' and Q', intersecting them at P'_x and Q'_y respectively. Then, given that the angle of shear is small, $\cos \psi$ and $\cos \phi$ are both very nearly equal to unity and so, to a very close approximation, $OP' = OP'_x$ and $OQ' = OQ'_y$.

Thus the extensional strain of the line of particles OP, called *the extensional strain in the* x *direction* and given the symbol ε_{xx}, is given by

$$\varepsilon_{xx} = \frac{OP'_x - OP}{OP} = \frac{PP'_x}{OP} \tag{2-3a}$$

Similarly, *the extensional strain in the* y *direction*, ε_{yy}, is given by

$$\varepsilon_{yy} = \frac{QQ'_y}{OQ} \tag{2-3b}$$

Again, since the strains are small, $\psi \simeq \tan \psi = P'P'_x/OP'_x$ and $OP'_x \simeq OP.†$ Thus the angle ψ, called *the shear strain in the* y *direction of a point on the* x *axis* and given the symbol ε_{yx}, is given by

$$\psi = \varepsilon_{yx} = \frac{P'P'_x}{OP} \tag{2-3c}$$

Similarly, *the shear strain in the* x *direction of a point on the* y *axis*, ε_{xy}, is given by

$$\phi = \varepsilon_{xy} = \frac{Q'Q'_y}{OQ} \tag{2-3d}$$

---

† The angle ψ is measured in radians. When angles are used to denote shear strains, radian measure will be used.

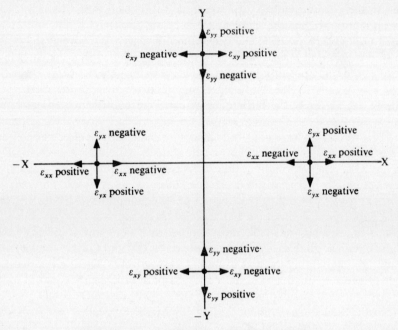

Fig. 2-4 Sign and notation convention for strain

We have used the symbol ε with two suffixes to denote strain. The usual convention in allocating these suffixes is as follows: *the first suffix denotes the direction of the component of displacement, the second suffix denotes the axis on which the particle lies.* The sign convention is as follows: *strains are positive if a particle moves in the direction of an axis which has the same sign as the axis on which the particle lies; they are negative if these signs are different.* This sign convention and notation is shown in Fig. 2-4, where, for example, if the particle lying on the x-negative axis is displaced in the positive y direction, the strain is $-\varepsilon_{yx}$.

From Eqns. (2-3) it would appear that we need four parameters to define strain. However, from Fig. 2-3

$$\phi + \psi = \tfrac{1}{2}\pi - \theta = \gamma_{xy}$$

where γ_{xy} is the angle of shear in the XY plane. Therefore, since the strains are small,

$$\varepsilon_{yx} + \varepsilon_{xy} = \gamma_{xy} \qquad (2\text{-}4)$$

Thus, ε_{yx} and ε_{xy} are not independent, since γ_{xy} is determined by the angle θ, and this is fixed for a given state of strain. By rigidly rotating the deformed body about O, ε_{yx} and ε_{xy} can be varied. The usual convention is to use this rigid rotation to make $\varepsilon_{yx} = \varepsilon_{xy}$, and, unless specifically stated otherwise, we will assume that this has been done in all cases in this chapter.

We therefore need three quantities to specify the strain in a sheet of material: two extensional strains and one shear strain. They are each defined by Eqns. (2-3). It is important to distinguish between the definition of γ_{xy}, the *angle of shear* [given in Eqn. (2-2c)], and ε_{xy}, the *shear strain* [given in Eqns. (2-3c) and (2-3d)], and it should be noted that

$$\varepsilon_{xy} = \tfrac{1}{2}\gamma_{xy}$$

It must again be emphasized that the definitions given in this section apply only to small strain. All future work, which will be developed from these, will therefore be similarly restricted. Where conclusions are also applicable to large strain, this will be stated, and when they can be easily developed to include large strains, this will be done.

2-4   Three-dimensional bodies

To extend the treatment to three-dimensional bodies, we need to consider the rotation and extension of three mutually perpendicular axes. In Fig. 2-5, three lines, OX, OY, OZ, are drawn in the material, and we define the directions in space of these lines as coordinate axes. Particles P, Q, R, one

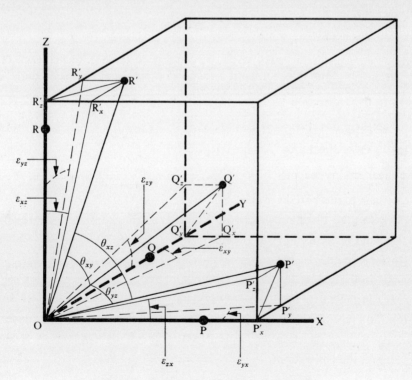

Fig. 2-5 Specification of strain of solid body

lying on each line, are marked. After deformation, the lines have rotated to the directions of OP′, OQ′, OR′, and the particles have been displaced to P′, Q′, R′. Again we assume that the lines remain straight after deformation, and that the strains are independent of the positions of the particles along the lines. From our work in two dimensions, we can see that six quantities are now necessary to specify the displacement of these points—three extensional strains and three angles of shear. The extensional strains are:

(a) in the x direction $\dfrac{\mathrm{OP}' - \mathrm{OP}}{\mathrm{OP}}$ (2-5a)

(b) in the y direction $\dfrac{\mathrm{OQ}' - \mathrm{OQ}}{\mathrm{OQ}}$ (2-5b)

(c) in the z direction $\dfrac{\mathrm{OR}' - \mathrm{OR}}{\mathrm{OR}}$ (2-5c)

The angles of shear are:

(a) in the XY plane $\quad \frac{1}{2}\pi - \theta_{xy}$ (2-5d)

(b) in the XZ plane $\quad \frac{1}{2}\pi - \theta_{xz}$ (2-5e)

(c) in the YZ plane $\quad \frac{1}{2}\pi - \theta_{yz}$ (2-5f)

As we are considering small strain, we can express these strains in terms of displacements along the coordinate axes. The extensional strains become

(a) in the x direction $\quad \varepsilon_{xx} = \mathrm{PP}'_x/\mathrm{OP}$ (2-6a)

(b) in the y direction $\quad \varepsilon_{yy} = \mathrm{QQ}'_y/\mathrm{OQ}$ (2-6b)

(c) in the z direction $\quad \varepsilon_{zz} = \mathrm{RR}'_z/\mathrm{OR}$ (2-6c)

The shear strains become

(a) for a particle on the x axis

   (i) in the y direction $\quad \varepsilon_{yx} = \mathrm{P}'_y\mathrm{P}'_x/\mathrm{OP}$ (2-6d)

   (ii) in the z direction $\quad \varepsilon_{zx} = \mathrm{P}'_z\mathrm{P}'_x/\mathrm{OP}$ (2-6e)

(b) for a particle on the y axis

   (i) in the x direction $\quad \varepsilon_{xy} = \mathrm{Q}'_x\mathrm{Q}'_y/\mathrm{OQ}$ (2-6f)

   (ii) in the z direction $\quad \varepsilon_{zy} = \mathrm{Q}'_z\mathrm{Q}'_y/\mathrm{OQ}$ (2-6g)

(c) for a particle on the z axis

   (i) in the x direction $\quad \varepsilon_{xz} = \mathrm{R}'_x\mathrm{R}'_z/\mathrm{OR}$ (2-6h)

   (ii) in the y direction $\quad \varepsilon_{yz} = \mathrm{R}'_y\mathrm{R}'_z/\mathrm{OR}$ (2-6i)

It would appear from Eqns. (2-6d)–(2-6i) that there are six shear strains.

However, as was the case with the two-dimensional lamina, these are not all independent. Since the shear strains are small

$$P'OQ' = P'_y OQ'_x \qquad R'OQ' = R'_y OQ'_z \qquad R'OP' = R'_x OP'_z$$

Hence

$$\varepsilon_{yx} + \varepsilon_{xy} = \tfrac{1}{2}\pi - \theta_{xy} = \gamma_{xy}$$

$$\varepsilon_{zy} + \varepsilon_{yz} = \tfrac{1}{2}\pi - \theta_{yz} = \gamma_{yz}$$

$$\varepsilon_{zx} + \varepsilon_{xz} = \tfrac{1}{2}\pi - \theta_{xz} = \gamma_{xz}$$

We can vary these strain components by rigidly rotating the body R'P'Q'O about O so that

$$\varepsilon_{yx} = \varepsilon_{xy} = \tfrac{1}{2}\gamma_{xy} \tag{2-7a}$$

$$\varepsilon_{zy} = \varepsilon_{yz} = \tfrac{1}{2}\gamma_{yz} \tag{2-7b}$$

$$\varepsilon_{zx} = \varepsilon_{xz} = \tfrac{1}{2}\gamma_{xz} \tag{2-7c}$$

2-5 Nature of strain

We see that *nine* parameters, of which *six* are independent, are necessary to define the strain in a piece of material. *Scalar* quantities can be completely defined by *one* number, and *vector* quantities require *three* independent numbers to specify them. Certain rules must be obeyed in the algebraic manipulation of these quantities, and the rules for scalars are different from those for vectors. However, since strain requires *six* independent numbers to specify it, it can be neither a scalar nor a vector quantity.

Scalars, vectors, and quantities such as strain are all given the general title *tensor quantities*. Scalars are zero rank tensors, vectors are first rank tensors, and quantities such as strain are second rank tensors. There are also tensors of higher rank. Strain is normally written as an array of nine numbers:

$$\begin{array}{ccc} \varepsilon_{xx} & \varepsilon_{xy} & \varepsilon_{xz} \\ \varepsilon_{yx} & \varepsilon_{yy} & \varepsilon_{yz} \\ \varepsilon_{zx} & \varepsilon_{zy} & \varepsilon_{zz} \end{array}$$

The first suffix is the same all along the rows, the second is the same down the columns. Also, by Eqn. (2-7), this array is symmetrical about the principal diagonal. (The principal diagonal is defined as that containing ε_{xx}, ε_{yy}, ε_{zz}.) Strain is therefore called a second rank symmetrical tensor quantity.

Like scalars and vectors, tensors of higher rank have their own algebraic rules. It is not necessary to go into these here, but one point should be emphasized—beware of using vector algebra when dealing with strain; in some situations this is a mistake which can easily be made by the unwary.

Large strain is also a second rank symmetric tensor quantity, but the strain components forming the elements of the tensor are defined differently from Eqns. (2-3).

While it is easy to visualize the physical nature of scalars and vectors (scalars are quantities having only magnitude, vectors have magnitude and direction), it is more difficult to visualize second (and higher rank) tensors. However, their nature can be illustrated, using strain as an example. In Fig. 2-6, P is a point in a body. After deformation, P will be displaced to P'.

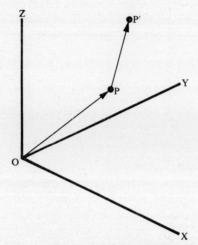

Fig. 2-6   Position and displacement vectors

The displacement will depend on both the state of strain and the position of P. If we denote this position by the vector **OP**, and the displacement by the vector **PP'**, then the strain is the quantity enabling the vector **PP'** to be calculated from the vector **OP**, i.e., the displacement and position vectors are related to the strain by the equation:

$$\text{displacement vector} = \text{strain} \times \text{position vector} \qquad (2\text{-}8)$$

Second rank tensors are quantities relating two vectors by an equation of this type.

From Eqns. (2-6), each strain component is a dimensionless ratio, which is often expressed as a percentage. For the small strain approximation to be valid, the numerical values of the components must be less than about 0·005, i.e., 0·5%.

2-6   Worked examples

1. Mutually perpendicular lines OX, OY are drawn in an undeformed sheet of material, and their directions define the coordinate axes. A particle P has the coordinates (2,0) and (2·005,0·002) before and after deformation, respectively. A particle Q has the coordinates (0,3) and (−0·003,2·998) before and after deformation, respectively. Calculate the state of strain and the angle of shear, assuming the strains are small.

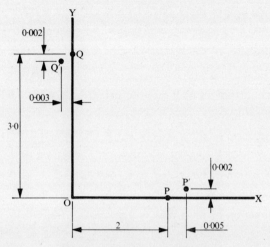

Fig. 2-7 Worked example 2-6(1)

The positions of the particles before and after deformation are shown in Fig. 2-7. From this figure, Eqns. (2-3), and Fig. 2-3,

$$\varepsilon_{xx} = 0{\cdot}005/2 = 2{\cdot}5 \times 10^{-3}$$
$$\varepsilon_{yy} = -0{\cdot}002/3 = -0{\cdot}6 \times 10^{-3}$$
$$\varepsilon_{yx} = 0{\cdot}002/2 = 1{\cdot}0 \times 10^{-3}$$
$$\varepsilon_{xy} = -0{\cdot}003/3 = -1{\cdot}0 \times 10^{-3}$$

From Eqn. (2-4), the angle of shear, γ_{xy}, is given by

$$\varepsilon_{xy} + \varepsilon_{yx} = \gamma_{xy}$$

Therefore

$$\gamma_{xy} = 0.$$

2. Mutually perpendicular lines OX, OY, OZ are drawn in an undeformed block of material, and their directions define the coordinate axes. The coordinates of particles P, Q, and R before deformation are (5,0,0), (0,2,0), and (0,0,3), respectively. Given that a state of small strain, $\varepsilon_{xx} = 4 \times 10^{-3}$, $\varepsilon_{yy} = 2 \times 10^{-3}$, $\varepsilon_{zz} = -1 \times 10^{-3}$, $\varepsilon_{xy} = 3 \times 10^{-3}$, $\varepsilon_{yz} = -5 \times 10^{-3}$, $\varepsilon_{xz} = 2 \times 10^{-3}$, ($\varepsilon_{xy} = \varepsilon_{yx}$, etc.), is applied, determine the coordinates of the particles after deformation.

From Eqn. (2-6), the displacement of P

(a) in the direction of the x axis $= \varepsilon_{xx} \times OP = 4 \times 10^{-3} \times 5$

$$= 2 \times 10^{-2}$$

(b) in the direction of the y axis $= \varepsilon_{yx} \times \mathrm{OP} = 3 \times 10^{-3} \times 5$
$$= 1\cdot5 \times 10^{-2}$$

(c) in the direction of the z axis $= \varepsilon_{zx} \times \mathrm{OP} = 2 \times 10^{-3} \times 5$
$$= 1 \times 10^{-2}$$

Therefore, the coordinates of P after deformation are $(5\cdot02, 0\cdot015, 0\cdot010)$.
The displacement of Q

(a) in the direction of the x axis $= \varepsilon_{xy} \times \mathrm{OQ} = 3 \times 10^{-3} \times 2$
$$= 6 \times 10^{-3}$$

(b) in the direction of the y axis $= \varepsilon_{yy} \times \mathrm{OQ} = 2 \times 10^{-3} \times 2$
$$= 4 \times 10^{-3}$$

(c) in the direction of the z axis $= \varepsilon_{zy} \times \mathrm{OQ} = -5 \times 10^{-3} \times 2$
$$= -1 \times 10^{-2}$$

Therefore, the coordinates of Q after deformation are $(0\cdot006, 2\cdot004, -0\cdot01)$.
The displacement of R

(a) in the direction of the x axis $= \varepsilon_{xz} \times \mathrm{OZ} = 2 \times 10^{-3} \times 3$
$$= 6 \times 10^{-3}$$

(b) in the direction of the y axis $= \varepsilon_{yz} \times \mathrm{OZ} = -5 \times 10^{-3} \times 3$
$$= -1\cdot5 \times 10^{-2}$$

(c) in the direction of the z axis $= \varepsilon_{zz} \times \mathrm{OZ} = -1 \times 10^{-3} \times 3$
$$= -3 \times 10^{-3}$$

Therefore, the coordinates of R after deformation are $(0\cdot006, -0\cdot015, 2\cdot997)$.

3. A sheet of material is in the form of a right-angled triangle whose mutually perpendicular edges are of length 3 cm and 4 cm. The directions of these edges are chosen as the coordinate axes, the edge of length 3 cm lying along OX. A state of strain, $\varepsilon_{xx} = 4 \times 10^{-3}, \varepsilon_{yy} = 2 \times 10^{-3}, \varepsilon_{xy} = \varepsilon_{yx} = 5 \times 10^{-3}$, is applied. Determine the extensional strain in the remaining edge of the triangle.

If the edges of the triangle after deformation are OX′ and OY′, and the angle between them is θ, then

$$(X'Y')^2 = (OX')^2 + (OY')^2 - 2(OX')(OY') \cos \theta$$

The angle of shear, γ_{xy}, is $2\varepsilon_{xy}$, and $\theta = \frac{1}{2}\pi - \gamma_{xy}$. Therefore

$$\cos \theta = \cos \left(\tfrac{1}{2}\pi - \gamma_{xy}\right) = \sin \gamma_{xy} = \gamma_{xy}$$

since γ_{xy} is very small.

Hence

$$(X'Y')^2 = (OX')^2 + (OY')^2 - 2\gamma_{xy}(OX')(OY')$$

where

$$OX' = OX + \varepsilon_{xx}OX$$

and

$$OY' = OY + \varepsilon_{yy}OY$$

Therefore

$$X'Y' = [OX^2(1 + \varepsilon_{xx})^2 + OY^2(1 + \varepsilon_{yy})^2$$
$$- 2\gamma_{xy}(OX)(OY)(1 + \varepsilon_{xx})(1 + \varepsilon_{yy})]^{1/2},$$

Similarly

$$XY = (OX^2 + OY^2)^{1/2}$$

The extensional strain in XY, ε, is given by

$$\varepsilon = \frac{X'Y' - XY}{XY}$$

Thus

$$\varepsilon = \left\{1 + \frac{2[\varepsilon_{xx}OX^2 + \varepsilon_{yy}OY^2 - \gamma_{xy}(OX)(OY)]}{OX^2 + OY^2}\right\}^{1/2} - 1$$

Since ε_{xx}, ε_{yy}, and γ_{xy} are sufficiently small, we can neglect second order terms. So, expanding by the binomial theorem, we get

$$\varepsilon = \frac{\varepsilon_{xx}OX^2 + \varepsilon_{yy}OY^2 - \gamma_{xy}(OX)(OY)}{OX^2 + OY^2}$$

Substituting the given values in this equation gives $\varepsilon = -0{\cdot}00208$.

Relevant exercises:† Nos. 2-1 to 2-3.

2-7 Uniform strain

We have not yet considered the implications of the two assumptions we made about deformation in Section 2-3—that the extensional strain along an axis is independent of the position of the particle, and that straight lines of particles used to define the axes remain straight after deformation.

The first of these assumptions implies that the extensional strain is uniform along the axis; the second has a similar implication for shear strain. If one of the lines becomes curved, then the angle of shear along that line must vary. Hence, the second assumption implies that shear strain is uniform along the axes.

However, if these two assumptions are correct, it does not necessarily mean that strain is uniform *at every point* in a sheet of material, for they are concerned with deformation *along the axes*. What are the necessary conditions for strain to be uniform throughout the sheet? Consider a sheet marked

† Exercises are given at the end of each chapter and students are recommended to work through the relevant questions before proceeding further.

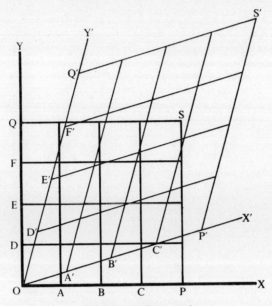

Fig. 2-8   Uniform strain

with a grid of small squares (Fig. 2-8). If, after deformation, all lines such as P'S' are straight and parallel to OY', then ε_{xy} must be uniform at all points, since it is this quantity which determines the changes in direction of lines initially parallel to OY. Similarly, if lines such as Q'S' are straight and parallel to OX', then ε_{yx} must be uniform at all points. It follows that, if the shear strains are uniform throughout the sheet, then lines of particles which are straight and parallel before deformation are straight and parallel afterwards.

Further, if the lengths OA, AB, BC, CP are all equal, and the extensional strain is uniform along the axis OX, then the lengths OA', A'B', B'C', C'P' must also all be equal. A similar conclusion applies to the intercepts OD, DE, EF, FQ along the axis OY.

Thus, if the shear strains ε_{xy} and ε_{yx} are uniform throughout the sheet, and the extensional strains ε_{xx} and ε_{yy} are uniform along the x and y axes, then a grid of congruent squares drawn in the sheet of material becomes a grid of congruent parallelograms after deformation, and the sheet is said to be in a state of *uniform strain*. It should be noted that it is not necessary to specify uniformity of extensional strain other than along the coordinate axes, since it follows from the above that the extensional strains are then uniform at all points. This can be shown in the following way. The condition of uniform shear strain ensures that lines parallel to PS before deformation are parallel to P'S' afterwards. Provided these lines mark off equal intercepts OA', A'B', B'C', C'P' along OX' (which is ensured by the condition of uniform extensional strain along this axis), they must also mark off equal intercepts along any other parallel line, i.e., the extensional strain must be uniform along any line parallel to OX and equal to ε_{xx}.

A similar definition of uniform strain applies to three-dimensional bodies. If the six shear strains are uniform throughout the body, and the three extensional strains are uniform along the coordinate axes, then congruent cubes drawn in the body before deformation become congruent parallelepipeds afterwards.

In this book we will be concerned only with states of uniform strain. For non-uniform strain, the definitions given in Eqns. (2-1), (2-2), and (2-3), and in Eqns. (2-5) and (2-6) must be replaced by their limiting values as the points P, Q, and R approach O. They then define the state of strain only at the point O.

2-8 Displacement of points not on coordinate axes

From our work in the previous section, we can now determine the displacement of a particle which does not lie on the coordinate axes.

Lines of particles forming the rectangle OQSP are marked in a sheet of material (Fig. 2-9). The directions in space of the lines OP and OQ are taken as the x and y axes, respectively, and the coordinates of S are (x,y). On deformation, the rectangle becomes the parallelogram OQ′S′P′. The lines of particles which coincided with the coordinate axes have moved to

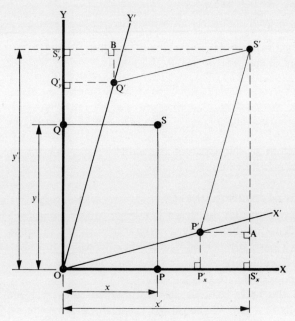

Fig. 2-9 Displacement during uniform strain

OX′ and OY′, and the coordinates of S′ are (x',y'). We need to find x' and y' in terms of x, y, and the state of strain.

Now

$$x' = OS_x' = OP + PP_x' + P_x'S_x' \qquad (2\text{-}9)$$

where
[from Eqn. (2-3a)]
$$OP = x \qquad \text{and} \qquad PP_x' = x\varepsilon_{xx}$$

and
[from Eqn. (2-3d)]
$$P_x'S_x' = P'A = Q'Q_y' = y\varepsilon_{xy}$$

Substituting these values in Eqn. (2-9) gives

$$x' = x(1 + \varepsilon_{xx}) + y\varepsilon_{xy} \qquad (2\text{-}10a)$$

Similarly, it can be shown that

$$y' = y(1 + \varepsilon_{yy}) + x\varepsilon_{yx} \qquad (2\text{-}10b)$$

The component of displacement in the x direction is $x' - x$, and that in the y direction is $y' - y$. These may be obtained from Eqns. (2-10) and written

$$x' - x = x\varepsilon_{xx} + y\varepsilon_{xy} \qquad (2\text{-}11a)$$
$$y' - y = x\varepsilon_{yx} + y\varepsilon_{yy} \qquad (2\text{-}11b)$$

The order in which the terms and suffixes occur in Eqns. (2-11) should be carefully noted. The *first* suffix remains the same *along the rows*, the *second* suffix remains the same *down the columns*. This enables the equations for the displacement of a point in a three-dimensional body to be written down by inspection. They are:

$$x' - x = x\varepsilon_{xx} + y\varepsilon_{xy} + z\varepsilon_{xz} \qquad (2\text{-}12a)$$
$$y' - y = x\varepsilon_{yx} + y\varepsilon_{yy} + z\varepsilon_{yz} \qquad (2\text{-}12b)$$
$$z' - z = x\varepsilon_{zx} + y\varepsilon_{zy} + z\varepsilon_{zz} \qquad (2\text{-}12c)$$

It is this pattern in the order of the suffixes which makes the method of notation particularly valuable.

2-9 Effect of rotation of axes

So far we have seen that the strain in a body can be completely specified by six independent numbers derived from the changes in length and angle between three mutually perpendicular lines drawn in it. The directions of these lines before deformation are taken as coordinate axes, and the values of strain obtained refer to this particular system of axes. Now suppose that we draw in the body another set of three mutually perpendicular lines, inclined to the first set. We could use the new set as a system of coordinate axes, and obtain numbers to specify the strain with respect to it. The numbers might differ from those specifying the strain with respect to the original system, even though the state of strain is unaltered. However, since the strain is completely specified by the first group of numbers, we should be able, knowing the angle between the two systems of axes, to calculate the strain components for the second system.

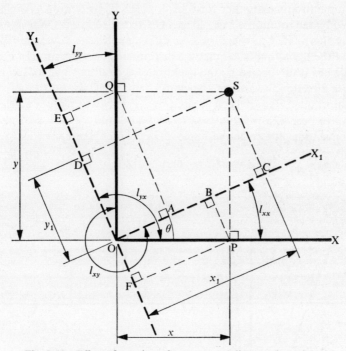

Fig. 2-10 Effect of rotation of axes on coordinates of a point

Consider a two-dimensional sheet having strain components ε_{xx}, ε_{yy}, and $\varepsilon_{xy} = \varepsilon_{yx}$ with respect to a system of axes OX, OY. For another system of axes OX_1, OY_1, inclined to the first at an angle θ, the state of strain is given by ε_{1xx}, ε_{1yy}, ε_{1xy}, and ε_{1yx}. (We are not justified in assuming that, because the shear strains are equal on one pair of axes, they are equal on any other pair.) We now have to derive equations which express ε_{1xx}, ε_{1xy}, ε_{1yy}, and ε_{1yx} in terms of ε_{xx}, ε_{yy}, ε_{xy}, and θ.

Before we can do this, we have to solve the following problem in coordinate geometry. Given that the coordinates of a point S are (x,y) with respect to axes OX, OY, what are its coordinates (x_1,y_1) with respect to axes OX_1, OY_1, inclined at an angle θ to OX, OY, in terms of x, y, and θ?

In Fig. 2-10, perpendiculars are drawn from Q, P, and S to A, B, and C on OX_1. Then

$$OC = OB + BC = x_1 \tag{2-13}$$

Now BC is the projection of SP on OX_1, and OA is the projection of OQ on OX_1. SP and OQ are equal and parallel, so OA is equal in length to BC. Substituting in Eqn. (2-13) gives

$$x_1 = OB + OA \qquad (2\text{-}14)$$

We now introduce another form of double suffix notation. Angle XOX_1 is formed by the rotation of the old axis OX to the new axis OX_1. We call the cosine of this angle l_{xx}, where the *first* suffix denotes the direction of the *new* axis, and the *second* suffix denotes the direction of the *old* axis. (In Fig. 2-10, the angle has been formed by the rotation of the old positive axis in an anti-clockwise direction towards the new positive axis. This direction of rotation is unimportant. Had it been formed by a clockwise rotation, the angle would have been $(2\pi - \theta)$ instead of θ and $\cos(2\pi - \theta) = \cos\theta$.) By the same convention, the cosine of angle YOX_1 is l_{xy}. Hence

$$OB = x l_{xx} \qquad \text{and} \qquad OA = y l_{xy}$$

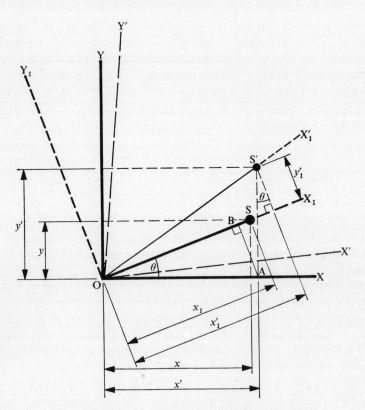

Fig 2-11   Effect of rotation of axes on specification of strain

Equation (2-14) can therefore be written

$$x_1 = xl_{xx} + yl_{xy} \qquad (2\text{-}15a)$$

Similarly, we can derive the equation

$$y_1 = xl_{yx} + yl_{yy} \qquad (2\text{-}15b)$$

The angles whose cosines are l_{yx} and l_{yy} are marked on Fig. 2-10 to assist the student to become familiar with this notation. He should note that the order in which the suffixes occur in these equations is exactly the same as in Eqns. (2-11), so that equations for the three-dimensional case can be written down by inspection. This form of notation offers little advantage while working in two dimensions, but is indispensible when working in three. It enables the mathematical techniques of matrix and tensor algebra to be applied to this type of problem, and is introduced at this stage so that the student may become familiar with it.

We can now solve the original problem of the effect of rotation of axes on the specification of strain. From Eqns. (2-11), the coordinates of a point before and after deformation can be expressed in terms of the state of strain. Using Eqns. (2-15) we can find the coordinates of the point with respect to the new system of axes. Using Eqns. (2-3), these transformed coordinates can be expressed in terms of the state of strain in the new system of axes.

It is convenient to consider the particle S in Fig. 2-11 lying on the OX_1 axis of the new system of coordinates. On deformation, the lines of particles coincident with the axes OX, OY, and OX_1 will be displaced to OX', OY', and OX_1', respectively, and the particle S will be displaced to S'. The coordinates (x_1', y_1') of S' in the OX_1, OY_1 system of axes are given by the equations

$$\varepsilon_{1xx} = (x_1'/x_1) - 1 \qquad (2\text{-}16a)$$

$$\varepsilon_{1yx} = (y_1'/x_1) \qquad (2\text{-}16b)$$

where x_1 is the distance along OX_1 of the particle S before deformation, and ε_{1xx} and ε_{1yx} are strains measured in the OX_1, OY_1 system of axes.

Also, from Fig. 2-11,

$$x = x_1 l_{xx} \qquad (2\text{-}17a)$$

$$y = x_1 l_{xy} \qquad (2\text{-}17b)$$

where (x, y) are the coordinates of S in the OX, OY system of axes. Applying Eqns. (2-15) to the coordinates of the point after deformation, and substituting for x' and y' from Eqns. (2-11) gives

$$x_1' = (x + x\varepsilon_{xx} + y\varepsilon_{xy})l_{xx} + (y + x\varepsilon_{yx} + y\varepsilon_{yy})l_{xy}$$
$$y_1' = (x + x\varepsilon_{xx} + y\varepsilon_{xy})l_{yx} + (y + x\varepsilon_{yx} + y\varepsilon_{yy})l_{yy}$$

Substituting from Eqns. (2-17) for x and y, rearranging, and using Eqns. (2-16) gives

$$1 + \varepsilon_{1xx} = (l_{xx} + l_{xx}\varepsilon_{xx} + l_{xy}\varepsilon_{xy})l_{xx} + (l_{xy} + l_{xx}\varepsilon_{yx} + l_{xy}\varepsilon_{yy})l_{xy} \quad (2\text{-}18a)$$

$$\varepsilon_{1yx} = (l_{xx} + l_{xx}\varepsilon_{xx} + l_{xy}\varepsilon_{xy})l_{yx} + (l_{xy} + l_{xx}\varepsilon_{yx} + l_{xy}\varepsilon_{yy})l_{yy} \quad (2\text{-}18b)$$

From Fig. 2-10

$$l_{xx} = \cos\theta \qquad l_{xy} = \sin\theta$$

$$l_{yy} = \cos\theta \qquad \text{and} \qquad l_{yx} = -\sin\theta$$

Substituting in Eqns. (2-18) and simplifying gives

$$\varepsilon_{1xx} = \varepsilon_{xx}\cos^2\theta + \varepsilon_{yy}\sin^2\theta + \varepsilon_{xy}\sin 2\theta \quad (2\text{-}19a)$$

$$\varepsilon_{1yx} = \varepsilon_{xy}\cos 2\theta + \tfrac{1}{2}(\varepsilon_{yy} - \varepsilon_{xx})\sin 2\theta \quad (2\text{-}19b)$$

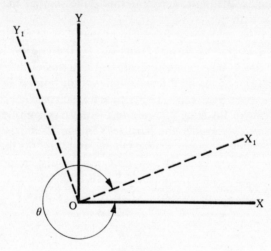

Fig. 2-12

To determine ε_{1yy} and ε_{1xy}, the substitution $\theta = \theta + \tfrac{1}{2}\pi$ can be made in Eqns. (2-19a) and (2-19b). This will give the strains measured on an axis at an angle of $\tfrac{1}{2}\pi$ to OX_1, which is OY_1. Substituting in Eqn. (2-19a) gives

$$\varepsilon_{1yy} = \varepsilon_{xx}\sin^2\theta + \varepsilon_{yy}\cos^2\theta - \varepsilon_{xy}\sin 2\theta \quad (2\text{-}19c)$$

Before substituting $\theta = \theta + \tfrac{1}{2}\pi$ in Eqn. (2-19b), it must be remembered that, not only does this have the effect of rotating OX_1 (Fig. 2-11) to OY_1, but OX_1' will also be rotated through a right angle. The resulting equation will therefore give a positive strain if a point on the positive OY_1 axis is displaced in the direction of the negative OX_1 axis. This is contrary to the sign convention, and therefore $-\varepsilon_{1xy}$ is substituted for ε_{1yx} in Eqn. (2-19b) when $\theta + \tfrac{1}{2}\pi$

is substituted for θ. This gives

$$\varepsilon_{1xy} = \varepsilon_{1yx}$$

In Fig. 2-11, the angle θ is formed by an anticlockwise rotation of the axes, and Eqns. (2-19) only relate to angles formed by rotation in this direction. We can show this by considering the result of clockwise rotation of the axes as in Fig. 2-12. The two sets of axes are the same as those in Fig. 2-11, and hence the direction cosines are unaltered. Equations (2-18) are therefore still valid. However, if the direction cosines are expressed in terms of the angle θ in Fig. 2-12, then

$$l_{xx} = \cos \theta \qquad l_{yy} = \cos \theta$$
$$l_{xy} = -\sin \theta \qquad \text{and} \qquad l_{yx} = \sin \theta$$

(whereas previously $l_{xy} = \sin \theta$ and $l_{yx} = -\sin \theta$, the other direction cosines being the same). Substituting these values in Eqns. (2-18) will *not* lead to Eqns. (2-19).

In deriving Eqns. (2-19) we have used the method which *must* be followed when working in three dimensions. The pattern in the order of suffixes enables us to use a form of shorthand notation which considerably reduces the number of algebraic terms to be written down. Using this notation we can solve the problem very elegantly for three dimensions.

However, if we are working in only two dimensions, a shorter derivation is possible, which will now be given. The student intending to go on to more advanced work is, however, recommended to become thoroughly familiar with the method already given; he will then only have to master the shorthand notation to understand the three-dimensional proof when he encounters it.

If, in Fig. 2-11, we draw the line AB, it will be seen that

$$x_1' = x' \cos \theta + y' \sin \theta$$

From Eqns. (2-11)

$$x_1' = (x + x\varepsilon_{xx} + y\varepsilon_{xy}) \cos \theta + (y + x\varepsilon_{yx} + y\varepsilon_{yy}) \sin \theta$$

Therefore, since $x_1 = x/\cos \theta$ and $y/x = \tan \theta$

$$\frac{x_1'}{x_1} = 1 + \varepsilon_{xy} \sin 2\theta + \varepsilon_{xx} \cos^2 \theta + \varepsilon_{yy} \sin^2 \theta$$

Now

$$\frac{x_1'}{x_1} - 1 = \varepsilon_{1xx}$$

and so

$$\varepsilon_{1xx} = \varepsilon_{xx} \cos^2 \theta + \varepsilon_{yy} \sin^2 \theta + \varepsilon_{xy} \sin 2\theta$$

which is the same as Eqn. (2-19a). Equations (2-19b) and (2-19c) can be derived similarly, and are left as exercises for the student.

2-10   Principal axes of strain and principal strains

Let us now consider Eqn. (2-19b), which describes the variation in shear strain as the axes are rotated. From this equation, the shear strain will be zero at an angle θ given by

$$\tan 2\theta = \frac{2\varepsilon_{xy}}{\varepsilon_{xx} - \varepsilon_{yy}} \tag{2-20}$$

Whatever the values of ε_{xy}, ε_{xx}, and ε_{yy}, the value of the right-hand side of this equation must lie between $\pm\infty$, i.e., there is always one value of θ between 0 and $\frac{1}{2}\pi$ which satisfies this equation.

In other words, whatever the state of strain in a sheet of material, one particular set of axes can always be found with respect to which the shear strain is zero. Since the shear strain is a measure of the change in angle between two lines of particles which were initially perpendicular, this means that there is always one pair of lines which are mutually perpendicular before and after straining.

Although we have reached this conclusion only from equations developed for small strains in a two-dimensional lamina, it can also be proved for three-dimensional bodies, and for large strains. Hence, it can be generally stated that, *for any strained body, there exists one set of mutually perpendicular lines of particles which were mutually perpendicular before deformation.* These are, of course, different for different states of strain, and are called the *principal axes of strain.* With respect to this particular set of axes, and to this set alone, the state of strain in a three-dimensional body can be completely defined by only three numbers—the extensional strains along the principal axes, known as the *principal strains.* To avoid confusion, we will denote principal strains by the use of a single suffix. When the principal axes are known, they are usually taken as the coordinate axes—for example, if the x axis is a principal axis, then ε_x is the principal strain along it.

The analysis of stress and strain is greatly simplified if the principal axes can be located. For small strain in a two-dimensional lamina we can do this using Eqn. (2-20). The same equation can be used for a three-dimensional problem, if we know one of the principal axes. Location of principal axes for the general case of three-dimensional strain, and for large strains, is a much more difficult problem and will not be dealt with here.

2-11   Mohr's circle construction

There is a graphical method of illustrating Eqns. (2-19), known as *Mohr's Circle Construction,* which provides a simple means of locating the principal axes and determining the principal strains for small, two-dimensional strain. On a sheet of graph paper a horizontal axis OX is drawn (Fig. 2-13), representing the extensional strains ε_{xx} and ε_{yy}; the vertical axis OY represents

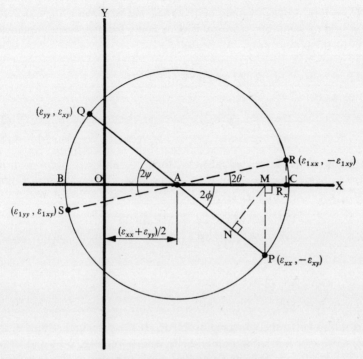

Fig. 2-13   Mohr's circle construction

the shear strain ε_{xy}. A point, A, is marked on OX at a distance equal to $\frac{1}{2}(\varepsilon_{xx} + \varepsilon_{yy})$ from the origin. Points P and Q are marked with coordinates $(\varepsilon_{xx}, -\varepsilon_{xy})$ and $(\varepsilon_{yy}, \varepsilon_{xy})$, respectively, and a circle is drawn, centre A, diameter PQ. The angle between PQ and the x axis, formed by an anticlockwise rotation of P to X, is 2ϕ. The circle intersects the x axis at B and C. It can be shown that principal axes of strain are at an angle ϕ (formed by anticlockwise rotation) to the axes used to define the strain. By rotating the axis through this angle we have the x principal axis, and it can also be shown that the principal strains ε_x and ε_y are given by the coordinates of C and B, respectively.

The validity of the construction can be verified as follows.

(a) Directions of principal axes

From Fig. 2-13

$$\tan 2\phi = MP/AM \qquad AM = \tfrac{1}{2}(\varepsilon_{xx} - \varepsilon_{yy}) \qquad \text{and} \qquad MP = \varepsilon_{xy}$$

Hence

$$\tan 2\phi = \frac{2\varepsilon_{xy}}{\varepsilon_{xx} - \varepsilon_{yy}}$$

Thus, from Eqn. (2-20), ϕ must be the angle between the principal axes and the axes used to define the strain, the angle being formed by an anticlockwise rotation of the latter axes.

(b) Magnitudes of principal strains

Rewriting Eqns. (2-19)

$$\varepsilon_{1xx} = \tfrac{1}{2}(\varepsilon_{xx} + \varepsilon_{yy}) + \tfrac{1}{2}(\varepsilon_{xx} - \varepsilon_{yy})\cos 2\theta + \varepsilon_{xy}\sin 2\theta \qquad (2\text{-}21a)$$

$$\varepsilon_{1yy} = \tfrac{1}{2}(\varepsilon_{xx} + \varepsilon_{yy}) - \tfrac{1}{2}(\varepsilon_{xx} - \varepsilon_{yy})\cos 2\theta - \varepsilon_{xy}\sin 2\theta \qquad (2\text{-}21b)$$

If θ in Eqns. (2-21) is put equal to ϕ, then ε_{1xx} and ε_{1yy} become principal strains, ε_{1xx} being labelled ε_x in the construction. Whence, substituting for $\tfrac{1}{2}(\varepsilon_{xx} + \varepsilon_{yy})$, etc., from Fig. 2-13 in Eqn. (2-21a),

$$\varepsilon_x = OA + AM\cos 2\phi + MP\sin 2\phi$$
$$= OA + AN + NP = OA + AP = OA + AC$$
$$= OC$$

Similarly, from Eqn. (2-21b),

$$\varepsilon_y = OA - AC = OB$$

We can also solve the converse problem. If the principal strains ε_x and ε_y are given, we can mark the points A, B, and C on Fig. 2-13 and draw the circle. The strains on axes at an angle θ anticlockwise to the principal axes are determined by drawing the diameter RS in Fig. 2-13. On the new axes, ε_{xx}, ε_{yy}, and ε_{xy} are given, respectively, by the x coordinates of R and S, and the y coordinate of S.

We can show this to be true by applying the condition that ε_{xx} and ε_{yy} are principal strains, i.e., $\varepsilon_{xx} = \varepsilon_x$, $\varepsilon_{yy} = \varepsilon_y$, $\varepsilon_{xy} = 0$, to Eqns. (2-21), giving

$$\varepsilon_{1xx} = \tfrac{1}{2}(\varepsilon_x + \varepsilon_y) + \tfrac{1}{2}(\varepsilon_x - \varepsilon_y)\cos 2\theta \qquad (2\text{-}22a)$$

$$\varepsilon_{1yy} = \tfrac{1}{2}(\varepsilon_x + \varepsilon_y) + \tfrac{1}{2}(\varepsilon_x - \varepsilon_y)\sin 2\theta \qquad (2\text{-}22b)$$

$$\varepsilon_{1xy} = \tfrac{1}{2}(\varepsilon_y - \varepsilon_x)\sin 2\theta \qquad (2\text{-}22c)$$

for the strains, with respect to axes at angle θ to the principal axes. Now in Fig. 2-13

$$OA = \tfrac{1}{2}(\varepsilon_x + \varepsilon_y)$$
$$AR = AC = \tfrac{1}{2}(\varepsilon_x - \varepsilon_y)$$

Therefore $AR_x = \tfrac{1}{2}(\varepsilon_x - \varepsilon_y)\cos 2\theta$ and from Eqn. (2-22a)

$$\varepsilon_{1xx} = OA + AR_x = OR_x$$

which is the x coordinate of R. The values obtained for ε_{1yy} and ε_{1xy} from the construction can similarly be shown to be correct.

It follows directly from the above that, if we are given the strains with respect to any one set of axes, enabling the points P, A, and Q to be marked and the circle to be drawn, then the strains with respect to any other set of axes inclined to the first at an angle ψ (measured anticlockwise) can be determined by drawing a diameter to the circle at an angle 2ψ to PQ (measured anticlockwise).

Certain other facts follow directly from this construction. We have shown that ε_{xx} and ε_{yy} are given by the x coordinates of the opposite ends of a diameter PQ. Hence, when ε_{xx} is a maximum, ε_{yy} is a minimum, and these maximum and minimum values are obtained when this diameter lies along the x axis, i.e., they are the principal strains. Similarly, the maximum numerical value of the shear strain is obtained on axes at 45° to the principal axes (i.e., when PQ is vertical in Fig. 2-13). This maximum shear strain is given by the radius of the circle, which, from Fig. 2-13, is

$$[\tfrac{1}{4}(\varepsilon_{xx} - \varepsilon_{yy})^2 + \varepsilon_{xy}^2]^{1/2}$$

where ε_{xx}, ε_{yy}, and ε_{xy} are the strains on any one set of axes. If the principal strains are known, the above expression becomes $\tfrac{1}{2}(\varepsilon_x - \varepsilon_y)$. Furthermore, at this angle, the extensional strains are equal and are given by the x coordinate of the centre of the circle, which is $\tfrac{1}{2}(\varepsilon_x + \varepsilon_y)$. Thus, if the principal strains are of equal magnitude, but of opposite sign, this coordinate will be zero and so, on axes at 45° to the principal axes, the extensional strains would be zero and the shear strain numerically equal to the principal strains. This particular kind of strain is known as *pure shear*.

2-12 Worked example

Coordinate axes are drawn in a sheet of material which is then subjected to a state of strain given by $\varepsilon_{xx} = 0\cdot005$, $\varepsilon_{yy} = 0\cdot003$, $\varepsilon_{xy} = \varepsilon_{yx} = -0\cdot002$. The origin of coordinates occupies the same position in space before and after deformation. Calculate:

(a) The coordinates after deformation of a particle initially at the point (1,3).

(b) The strains on axes at an angle of 30° anticlockwise to the given axes.

(c) The directions of the principal axes and the magnitudes of the principal strains.

(d) The directions of the axes on which shear strain is a maximum, and the strain components on these axes.

(e) The change in angle between the line of particles lying along the original x axis and a line at 30° to it before straining.

(a) From Eqns. (2-11), if (x',y') are the coordinates of the point after deformation,

$$x' = (1 + 0\cdot005) - (3 \times 0\cdot002) = 0\cdot999$$
$$y' = (3 - 0\cdot002) + (3 \times 0\cdot003) = 3\cdot007$$

(b) From Eqns. (2-19), if OX_1 is the new axis at 30° anticlockwise to OX, and OY_1 the new y axis,

$$\varepsilon_{1xx} = 0\cdot005 \cos^2 30° + 0\cdot003 \sin^2 30° - 0\cdot002 \sin 60° = 0\cdot00277$$
$$\varepsilon_{1yy} = 0\cdot005 \sin^2 30° + 0\cdot003 \cos^2 30° + 0\cdot002 \sin 60° = 0\cdot00523$$
$$\varepsilon_{1yx} = \varepsilon_{1xy} = -0\cdot002 \cos 60° + \tfrac{1}{2}(0\cdot003 - 0\cdot005) \sin 60° = -0\cdot001865$$

(c) From Eqn. (2-20), the principal axes are at an angle θ anticlockwise to the coordinate axes, given by

$$\tan 2\theta = \frac{-2(0\cdot002)}{(0\cdot005 - 0\cdot003)} = -2$$

Therefore

$$\theta = 58°17'$$

If the axis at 58°17′ anticlockwise from the x axis is called the x principal axis, then from Eqns. (2-19),

$$\varepsilon_x = 0\cdot005 \cos^2 58°17' + 0\cdot003 \sin^2 58°17' - 0\cdot002 \sin 116°34' = 0\cdot00176$$
$$\varepsilon_y = 0\cdot005 \sin^2 58°17' + 0\cdot003 \cos^2 58°17' + 0\cdot002 \sin 116°34' = 0\cdot00624$$

(d) The shear strain is a maximum on axes at 45° to the principal axes, i.e., on axes at an angle of 13°17′ anticlockwise to the coordinate axes.

The shear strain is $\qquad \tfrac{1}{2}(\varepsilon_x - \varepsilon_y) = -0\cdot00224$

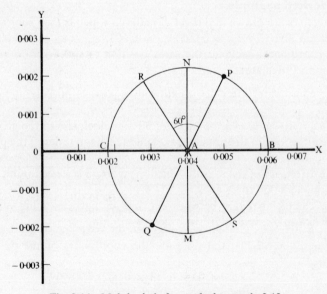

Fig. 2-14 Mohr's circle for worked example 2-12

The extensional strains, which are equal, are

$$\tfrac{1}{2}(\varepsilon_x + \varepsilon_y) = 0 \cdot 004$$

(e) Since $\varepsilon_{xy} = -0 \cdot 002$, the line of particles initially along the positive x axis will rotate through $0 \cdot 002$ radians towards the negative y axis. (For small strain, the shear strain is equal to the angle of rotation of the axis in radians. Since the shear strain is negative, a positive axis is displaced towards a negative.)

From the solution of part (b), the line of particles at $30°$ to the positive x axis will rotate through $0 \cdot 001865$ radians towards the negative y axis. Hence, the angle between the two lines will be increased by $0 \cdot 000135$ radians.

It is instructive to draw Mohr's circle for this problem (Fig. 2-14). The points P, A, and Q are marked according to the instructions in Section 2-11. Then the strains on axes at $30°$ anticlockwise to the given axes are obtained by drawing the line RS, the coordinates of R being $(\varepsilon_{1xx}, -\varepsilon_{1xy})$ and of S$(\varepsilon_{1yy}, \varepsilon_{1xy})$. The x principal axis is formed by rotating the x axis anticlockwise through $\tfrac{1}{2}\angle$ PAC, and ε_x and ε_y are given by the x coordinates of C and B respectively. If the given axes are rotated anticlockwise through $\tfrac{1}{2}\angle$ PAN, axes will be obtained on which shear strain is a maximum, and on these axes ε_{xx}, ε_{yy}, and ε_{xy} are given by the coordinates of N$(\varepsilon_{xx}, -\varepsilon_{xy})$ and M$(\varepsilon_{yy}, \varepsilon_{xy})$.

Relevant exercises: Nos. 2-4 to 2-10.

2-13   The strain ellipsoid

We have now seen that, for small strains in a two-dimensional sheet, the state of strain can be completely specified by two principal strains. It has also been stated that a similar result holds for three-dimensional bodies and for large strains. We will now calculate the change on straining in the position of the particles of a body with respect to the principal axes, and in terms of the principal strains. We could use Eqns. (2-19) for this, but we can derive the answer quite easily from first principles.

Consider a particle at P (Fig. 2-15) in a sheet of material to which uniform strains ε_x and ε_y are applied with respect to the principal axes OX and OY. The particle will be displaced to P', and the lines of particles PP$_x$ and PP$_y$ will be displaced to P'P$'_x$ and P'P$'_y$, respectively. Since OX and OY are the principal axes, the lines of particles defining their directions will remain mutually perpendicular after deformation. Hence, the particles forming the rectangle OP$_x$PP$_y$ will, on deformation, form the rectangle OP$'_x$P'P$'_y$ (Section 2-7).

From Eqns. (2-3)

$$x' = x(1 + \varepsilon_x) \tag{2-23a}$$

$$v' = y(1 + \varepsilon_y) \tag{2-23b}$$

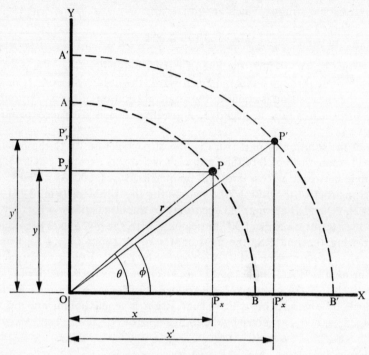

Fig. 2-15   The strain ellipsoid

From Fig. 2-15

$$x = r \cos \theta \quad \text{and} \quad y = r \sin \theta$$

Substituting in Eqns. (2-23) gives

$$x' = r(1 + \varepsilon_x) \cos \theta \tag{2-24a}$$

$$y' = r(1 + \varepsilon_y) \sin \theta \tag{2-24b}$$

Now suppose that P lies on the circular arc AB, then x' and y' are the coordinates after deformation of particles lying initially on a circular arc. In this case, r is a constant, and θ can be eliminated between Eqns. (2-24a) and (2-24b) giving

$$\frac{x'^2}{(1 + \varepsilon_x)^2} + \frac{y'^2}{(1 + \varepsilon_y)^2} = r^2 \tag{2-25}$$

as the equation of the arc A'B' after deformation. This represents an ellipse whose major and minor axes coincide with the coordinate axes, which are the principal axes of strain. If r is made equal to unity, the lengths of the major and minor axes are equal to the principal strains added to unity. A similar result can be proved for three-dimensional bodies, and so we can state that a sphere drawn in a body becomes an ellipsoid after straining, the principal axes of which coincide with the principal axes of strain. If the sphere is of unit radius, the lengths of the principal axes are equal to unity plus the principal strains. This ellipsoid is known as the *strain ellipsoid*; it provides a convenient illustration of the changes in shape and size of a body on straining.

Referring again to Fig. 2-15, if ϕ is the angle between the radius vector to the particle and the x principal axis after deformation, then

$$\tan \phi = \frac{y'}{x'} = \frac{1 + \varepsilon_y}{1 + \varepsilon_x} \tan \theta \qquad (2\text{-}26)$$

This equation gives the rotation, on straining, of a line drawn in the body.

If r' is the length after deformation of a radius vector which was of unit length before deformation, then, from Fig. 2-15 and Eqn. (2-25),

$$\frac{\cos^2 \phi}{(1 + \varepsilon_x)^2} + \frac{\sin^2 \phi}{(1 + \varepsilon_y)^2} = \frac{1}{r'^2} \qquad (2\text{-}27)$$

which is the equation of the strain ellipsoid in polar form.

From Eqns. (2-26) and (2-27) we can calculate the displacement on straining of any particle. Equation (2-27) gives the change in length of the line joining the particle to the origin of coordinates and Eqn. (2-26) gives the rotation of the line. Furthermore, we can use these equations to calculate the strain on axes inclined to the principal axes—the extensional strain from Eqn. (2-27), and the shear strains from Eqn. (2-26).

Our conclusions on the strain ellipsoid have been derived from Eqns. (2-3) and so only apply to small strain. It can, however, be shown that a sphere still becomes an ellipsoid after large straining and that the principal axes of this ellipsoid are the principal axes of strain.

For large strain, it is convenient to define the *extension ratio*, which is the ratio of the length of a line of particles in the deformed body to its length before deformation.

If the sphere drawn in the unstrained body has unit radius, the length of a radius vector of the strain ellipsoid will give the extension ratio of the line of particles lying along this vector. The lengths of the principal axes are called the *principal extension ratios*.

In previous sections, strain has been specified as a group of numbers, and whilst this is convenient mathematically it gives no visual picture of the state of strain. The strain ellipsoid, however, is a geometrical concept and so enables the state of strain to be visualized. If one imagines a sphere of unit radius to be drawn in the body before straining, the strain ellipsoid gives the size and shape assumed by this sphere after straining. The remaining sections of this chapter will develop the application of this concept to the analysis of strain.

2-14   Strain ellipsoid for equal principal strains

Suppose a sheet of material is strained so that the principal strains are equal. Then

$$\varepsilon_x = \varepsilon_y = \varepsilon$$

Substituting this in Eqn. (2-25) gives the equation of the strain ellipse for this type of strain as

$$x^2 + y^2 = (1 + \varepsilon)^2 \tag{2-28}$$

if r equals unity. This is the equation of a circle of radius $1 + \varepsilon$.

We can see that all particles equally distant from the origin of coordinates before deformation must still be equally distant afterwards, whatever the orientation in the sheet of the line joining the particle to the origin (in other words, the extensional strain in all radius vectors is equal). Also, since the strain ellipse is a circle, the principal axes, which are the major and minor axes of this ellipse, are not uniquely determined. Hence any pair of mutually perpendicular axes can be chosen as principal axes.

This second conclusion can also be deduced in another way. From Eqn. (2-26), if the principal strains are equal, $\phi = \theta$. This means that when the sheet is deformed the rotation relative to the principal axes of all lines of particles, whatever their orientation, will be zero. It therefore follows that the angle between any pair of lines will remain unchanged and so the angle of shear between any pair of axes will be zero. Since principal axes are defined as those for which the angle of shear (or the shear strains) is zero, it follows that any pair of mutually perpendicular axes can be chosen as principal axes.

A similar result holds for three-dimensional bodies in which the three principal strains are equal. In this case *a sphere drawn in the body becomes a sphere of different radius on straining and the three shear strains are all zero, whatever axes are used to measure them, i.e., the size, or volume, of the sphere changes, but not its shape.*

This special kind of strain is called *dilatation*, and since only changes in volume are involved it can be specified by one number only, called the *dilatational strain*, Δ, and defined as

$$\Delta = \frac{\text{change in volume}}{\text{original volume}}$$

Another quantity which is useful in defining the dilatational strain is the principal strain in dilatation, ε_d.

From Eqn. (2-28), if a sphere of unit radius is drawn in an unstrained body, after dilatation its radius is $1 + \varepsilon_d$. Therefore, from the definition of dilatational strain,

$$\Delta = \frac{\frac{4}{3}\pi(1 + \varepsilon_d)^3 - \frac{4}{3}\pi}{\frac{4}{3}\pi}$$

If the strains are small, second and higher order terms in ε_d can be neglected, giving

$$\Delta = 3\varepsilon_d \qquad (2\text{-}29)$$

It is instructive to consider the relative movement of particles of the material during dilatational strain. In Fig. 2-16 the solid circles represent particles lying on a square lattice before deformation. The body is then subjected to a dilatational strain with OX and OY as principal axes, the particle O occupying the same position in space before and after deformation. From the preceding discussion, the directions in space of the lines of particles OY, OA, OB, OC, etc., will not change on deformation and the extensional strains along all such lines will be the same. So we can locate each particle after deformation. The positions are indicated by open circles, and we can see that after deformation the particles remain on a square lattice, larger than the original one.

Hence, in dilatational strain, the distance between particles will increase but the angles between lines of particles will not change. On the atomic scale, this means that, if the atoms of the material are packed in a regular crystalline pattern, dilatational strain will increase their spacing but will not change the angle between the crystal axes. This conclusion is important for interpreting the elastic properties of matter in terms of molecular structure.

All the conclusions in this section apply equally to large and small strains except Eqn. (2-29), which was derived using a small strain approximation.

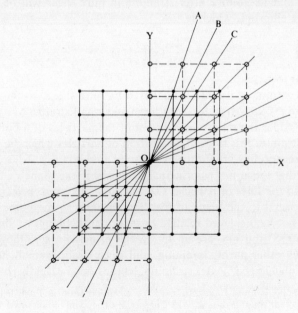

Fig. 2-16 Relative displacement of particles in dilatation

2-15   Strain ellipsoid for pure shear

In Section 2-11 we defined pure shear as the state of strain which occurs in a two-dimensional sheet when the principal strains are of equal magnitude but of opposite sign. As a corollary, it is shown that, if the strain is referred to axes at 45° to the principal axes, the extensional strains are zero and the shear strain numerically equal to the principal strain. This state of strain is not necessarily restricted to two-dimensional sheets. It can occur in a three-dimensional body if the third principal strain is zero. According to this definition then, pure shear is the state of strain represented by the principal strains ε, $-\varepsilon$, 0.

It is important to realize that this definition, and its corollary, is true only for small strain. The more general definition, true for large or small strain, states that pure shear is the state occurring when the principal extension ratios of the strain ellipsoid are α,† $1/\alpha$, 1. If strain is small, Eqn. (2-3) can be used to relate the extensional strain to the extension ratio α, giving

$$\alpha = 1 + \varepsilon \quad \text{and} \quad 1/\alpha = 1/(1 + \varepsilon)$$

Powers of ε higher than the first can be neglected, and so, by a binomial expansion,

$$1/(1 + \varepsilon) = 1 - \varepsilon$$

The two definitions are therefore equivalent at small strain.

The volume of a sphere of unit radius drawn in a body before straining will be $\frac{4}{3}\pi$, and its volume after straining in pure shear will be the volume of the strain ellipsoid, which is

$$\frac{4\pi}{3} (\alpha) \left(\frac{1}{\alpha}\right) (1)$$

and is equal to $\frac{4}{3}\pi$.

Thus the volume of a body is unaltered by pure shear, so the dilatational strain is zero. This is true for both large and small strain.

Let us consider the relative movement of the particles of a body subjected to a small strain of this type. The position of particles which lie on a square lattice before straining are shown by the solid circles in Fig. 2-17. The body is strained so that the non-zero principal strains lie in the plane of the diagram and at 45° to the lines of particles OX and OY. The zero principal strain is therefore normal to the plane of the diagram, and so the plane of particles shown will not be displaced relative to other parallel planes. Since the lines of particles OX and OY are at 45° to the principal axes, they will rotate towards each other during straining, and since strain is small, their lengths will remain unchanged. After straining, let the body be rigidly rotated so that

† The symbol λ is more frequently used for extension ratio in English texts, but this symbol is used for another quantity in Chapter 4. The symbol α is used in the United States.

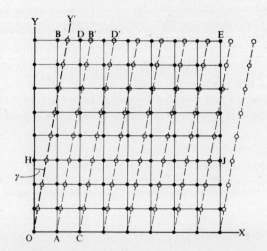

Fig. 2-17 Relative displacement of particles in shear

OX occupies the same direction in space as it did before straining. Since this line is unextended, each particle on it will then occupy its original position in space. The straining and rigid body movement will have caused OY to rotate through an angle of shear γ to OY′, where γ is equal to twice the shear strain. Since the strain is uniform, all other lines of particles such as AB and CD, which were initially parallel to OY, will be rotated through the angle γ to AB′, CD′, etc., parallel to OY′. Hence the array of particles form a parallelogram lattice.

We are faced now with an apparent contradiction. By our definition of pure shear, the lines of particles OY′, etc., will be unchanged in length, so the perpendicular distance between YE and OX must decrease slightly on straining. But we have seen that, in pure shear, the volume of a body does not alter, so this perpendicular distance cannot change.

However, the definition is an approximation valid only at small strains, for which the change in length of OY is negligible. At large strains, lines at 45° to the principal axes change in length, and so the difficulty does not arise.

The deformation shown in Fig. 2-17 is a pure shear strain plus a rigid body rotation, so that instead of the directions in space of the principal axes remaining unchanged on straining (as they do in pure shear), the directions of lines of particles at 45° to these axes remain unchanged. Such a deformation is called *simple shear*. This definition is, however, valid only for small strain. At large strains, the rigid body rotation is such that the direction in space of a line unextended by straining remains the same as that before straining. This line is at 45° to the principal axes only for small strains.

Usually, simple shear is defined as the deformation shown in Fig. 2-17, in which the length OX and the perpendicular distance between OX and YE both remain constant (as will the volume). This definition is true for small or large strain.

Pure and simple shear differ only by a rigid body rotation. In pure shear, the directions of the principal axes remain fixed, whereas in simple shear the direction of the line OX remains fixed.

In pure or simple shear, parallel lines of particles (or, in three-dimensional bodies, sheets of particles) slide over each other, their perpendicular separation remaining constant. On the atomic scale this means that, if the atoms are packed in a regular crystalline pattern, then, in pure or simple shear, deformation occurs by parallel sheets of atoms sliding over each other. This is another important conclusion for interpreting elastic properties in terms of molecular structure.

2-16 Deviatoric strain

By definition, pure shear is a type of deformation which can take place only in two dimensions; the third principal strain must be zero. For this type of deformation the dilatational strain is zero, but this is not the only way to achieve zero dilatation; it can be caused by other types of strain for which the third principal strain is not necessarily zero. Any strain which causes zero dilatation is called a *deviatoric* strain. Since *dilatation* changes only the volume of a body, a *deviatoric* strain must change only its shape. We will see later that it can be represented by two pure shears applied simultaneously, each to a different pair of principal axes.

2-17 Analysis of strain into dilatational and deviatoric components

Two special types of strain have now been defined; dilatational and deviatoric. We can show that any type of strain can be resolved into these components, as follows.

In Fig. 2-18, (a) represents a sphere of unit radius drawn in a body before straining, and (c) represents the ellipsoid into which this sphere is transformed

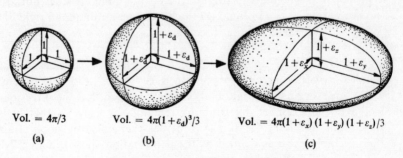

Vol. $= 4\pi/3$ Vol. $= 4\pi(1+\varepsilon_d)^3/3$ Vol. $= 4\pi(1+\varepsilon_x)(1+\varepsilon_y)(1+\varepsilon_z)/3$

(a)   (b)   (c)

Fig. 2-18 Strain applied in two stages

by straining, the principal strains being ε_x, ε_y, ε_z. Let this straining take place in two stages:

In stage 1, a dilatational strain ε_d is applied to produce a sphere of larger radius, as in Fig. 2-18b. Let the magnitude of this dilatational strain be such that the volume of the sphere in (b) is equal to that of the ellipsoid in (c).
In stage 2, a deviatoric strain is applied to the sphere in (b) to produce the ellipsoid in (c).

The vector representing the displacement of any particle during the strain from (a) to (c) can be regarded as the sum of two component vectors—the displacement during the dilatational strain from (a) to (b) and the displacement during the deviatoric strain from (b) to (c). However, although these displacements can be regarded as components of the total displacement, the strains can only be regarded as components of the total strain if they are small. This can be shown as follows. For a given particle, let the strain from (a) to (b) be A, then from Eqn. (2-8) and Fig. 2-6

$$\text{displacement vector (a)} \rightarrow \text{(b)} = A \times \text{position vector (a)}$$

Similarly, for the same particle, let the strain from (b) to (c) be B, then:

$$\text{displacement vector (b)} \rightarrow \text{(c)} = B \times \text{position vector (b)}$$

The terms 'position vector (a)' and 'position vector (b)' refer to the vector from the origin of coordinates to the particle at stages (a) and (b) respectively.

Now suppose that, instead of applying strains A and B successively to the particle, we applied a strain $A + B$ in one operation. The vector representing the displacement would then be given by

$$(A + B) \times \text{position vector (a)}$$

which may be written as the vector sum

$$[A \times \text{position vector (a)}] + [B \times \text{position vector (a)}]$$

The first of these vectors is the displacement vector (a) $\rightarrow$ (b), but the second is *not* the displacement vector (b) $\rightarrow$ (c).

So the result of applying strains A and B successively to a body, as in Fig. 2-18, is not the same as that of applying strain $A + B$ in one operation, i.e., strain is not a *superposable* quantity. Superposability is a necessary condition for an operation to be broken down into component parts. Otherwise, not only will the effect of performing the operation in one stage differ from the effect of performing its components successively, but the order in which the components are performed will also be important. However, it also follows from the above argument that, if the position vector (a) is not significantly different from the position vector (b) (i.e., the strain is small), then strain becomes a superposable quantity. We can show this by a simple example.

Suppose a thin filament of unit length is given a strain ε_1 along its axis. Its length will become $1 + \varepsilon_1$. If it is now given a strain ε_2, its length will become [by Eqn. (2-3)]

$$(1 + \varepsilon_1)(1 + \varepsilon_2) \quad \text{or} \quad 1 + \varepsilon_1 + \varepsilon_2 + \varepsilon_1\varepsilon_2.$$

However, if a strain $\varepsilon_1 + \varepsilon_2$ was applied in one operation, its length would become $1 + \varepsilon_1 + \varepsilon_2$. Hence, these strains are not superposable unless $\varepsilon_1\varepsilon_2$ is negligible compared with $\varepsilon_1 + \varepsilon_2$, which is only the case if the strains are small.

So we can say that, provided the strains are small, a dilatational strain A and a deviatoric strain B can be regarded as components of a total strain C, so that

$$A + B = C$$

and the effect of applying strains A and B successively is the same as that of applying C in one operation.

In this expression A, B, and C are second rank symmetric tensors, and the addition must be done as follows. Suppose the elements of one tensor referred to a particular set of coordinate axes are ε_{11}, ε_{12}, etc., and the elements of another referred to *the same coordinate axes* are ε'_{11}, ε'_{12}, etc. It can be shown that the elements of the tensor sum of these two are $\varepsilon_{11} + \varepsilon'_{11}$, $\varepsilon_{12} + \varepsilon'_{12}$, etc., referred to the same coordinate axes. In this book we are concerned only with adding states of strain whose principal axes coincide, and which are described by their principal values. It follows from the above rule for tensor addition that the principal axes of the resultant strain will coincide with those of its components, and the resultant principal strain along any axis will be equal to the sum of the principal values of the components along the same axis. This is shown in Fig. 2-19, in which the heads of the arrows of unit length drawn from the origin represent the points where a sphere of unit

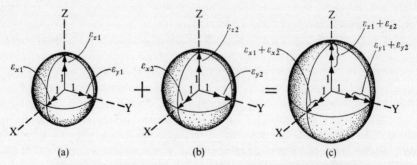

Fig. 2-19 Addition of strains with the same principal directions

radius drawn in the undeformed body would intersect the principal axes. The lengths of the short arrows, labelled ε_{x1}, etc., therefore represent the principal strains. Figure 2-19(a) represents the strain ellipsoid and principal strains resulting from applying strain A, and (b) represents strain B, with coincident principal axes. In (c), the principal strains are added to give the strain ellipsoid resulting from applying strain $A + B$.

To return now to Fig. 2-18, we have seen that if the strain applied between stages (b) and (c) is deviatoric, then the volume of the sphere in (b) equals that of the ellipsoid in (c). That is,

$$\tfrac{4}{3}\pi(1 + \varepsilon_d)^3 = \tfrac{4}{3}\pi(1 + \varepsilon_x)(1 + \varepsilon_y)(1 + \varepsilon_z)$$

Neglecting second and higher order terms in ε gives

$$\varepsilon_d = \tfrac{1}{3}(\varepsilon_x + \varepsilon_y + \varepsilon_z) \tag{2-30}$$

and, from Eqn. (2-29),

$$\varDelta = \varepsilon_x + \varepsilon_y + \varepsilon_z \tag{2-31}$$

$\varDelta$ in Eqn. (2-31) is the dilatational strain applied in the stage (a) to (b); ε_d is the principal strain in dilatation applied in this stage. Since the stage (b) to (c) is entirely deviatoric, these quantities are alternative means of representing the dilatational component of the total strain represented by the principal strains ε_x, ε_y, ε_z.

It is shown in Section 2-14 that in dilatation any set of axes can be chosen as principal axes. If, therefore, principal axes are chosen that coincide with those of the final state of strain, then the principal values of the deviatoric strain component can be found by subtracting ε_d from ε_x, ε_y, and ε_z. The symbols ε_{xd}, ε_{yd}, and ε_{zd} are used to denote the deviatoric components obtained in this way. Then, from Eqn. (2-30),

$$\varepsilon_{xd} = \tfrac{1}{3}(2\varepsilon_x - \varepsilon_y - \varepsilon_z) \tag{2-32a}$$

$$\varepsilon_{yd} = \tfrac{1}{3}(-\varepsilon_x + 2\varepsilon_y - \varepsilon_z) \tag{2-32b}$$

$$\varepsilon_{zd} = \tfrac{1}{3}(-\varepsilon_x - \varepsilon_y + 2\varepsilon_z) \tag{2-32c}$$

If this is a deviatoric strain, then the dilatation occurring when it is applied should be zero. From Eqn. (2-31), the dilatation is equal to the sum of the principal strains, and can be found by adding together Eqns. (2-32). This gives

$$\varepsilon_{xd} + \varepsilon_{yd} + \varepsilon_{zd} = 0 \tag{2-33}$$

showing that the dilatation is, as it should be, zero.

The deviatoric component of the strain can be further analysed into pure shears. To prove this, we need, first of all, to show that a deviatoric strain can be produced by simultaneously applying three pure shears, one in each of the principal planes. (A principal plane is the plane containing two

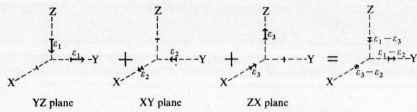

YZ plane   XY plane   ZX plane

Fig. 2-20   Addition of three pure shears in principal planes

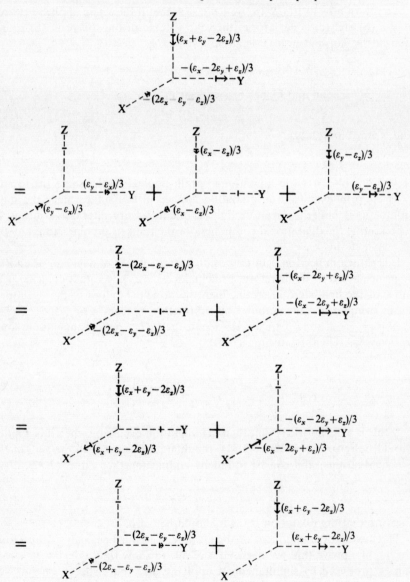

Fig. 2-21   Analysis of deviatoric strain into pure shears

principal axes.) We use the definition of small pure shear from Section 2-15 (i.e., it consists of two principal strains of equal magnitude but opposite sign, the third principal strain being zero). Three such shears are shown in Fig. 2-20,† and superimposing them gives the principal values of the resultant strain as

$$\varepsilon_x = \varepsilon_2 - \varepsilon_3 \qquad (2\text{-}34a)$$

$$\varepsilon_y = \varepsilon_1 - \varepsilon_2 \qquad (2\text{-}34b)$$

$$\varepsilon_z = \varepsilon_3 - \varepsilon_1 \qquad (2\text{-}34c)$$

(ε_x, ε_y, and ε_z are positive when outward acting.) Adding these equations, we obtain

$$\varepsilon_x + \varepsilon_y + \varepsilon_z = 0$$

which is the condition for deviatoric strain.

So we can write ε_x, ε_y, and ε_z in Eqns. (2-34) as ε_{xd}, ε_{yd}, and ε_{zd}, respectively, and substitute in Eqns. (2-32), which can then be solved for ε_1, ε_2, and ε_3. But there is no unique solution. Four different possible solutions are shown in Fig. 2-21, which all represent different ways of analysing the deviatoric component of the total strain into pure shears.

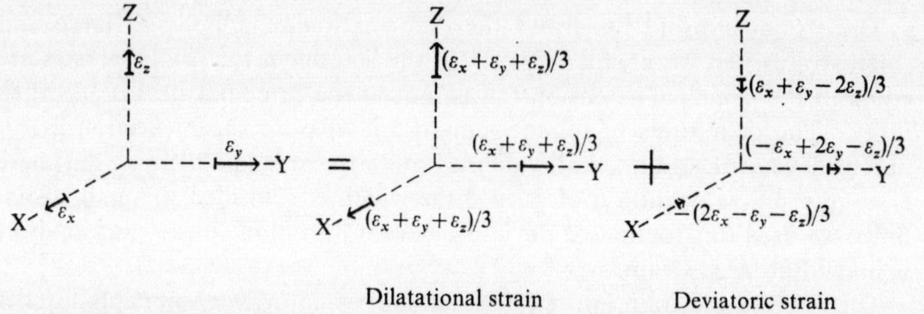

Dilatational strain   Deviatoric strain

Fig. 2-22   Analysis of strain into dilatational and deviatoric components

Figure 2-22 summarizes the conclusions from this section. The figure shows the analysis of a general state of strain ε_x, ε_y, ε_z into dilatational and deviatoric components. The component of each principal strain due to dilatation is given by Eqn. (2-30), and that due to deviatoric strain by Eqn. (2-32). The deviatoric strain can be further broken down into pure shears, but this cannot be done uniquely, and various possible methods are illustrated in Fig. 2-21.

† This figure, and subsequent ones, are to be interpreted similarly to Fig. 2-19, the strain ellipsoid and unit arrows being omitted for simplicity. The arrows remaining represent the principal strains, as did the equivalents in Fig. 2-19, and the lengths of those along the same principal axis can be added to give the resultant principal strain.

2-18   Restriction to small strain

In previous sections we have been concerned with states of small strain and only at a few points have we extended the discussion to include large strain. The restriction to small strain is very important and we will now consider its implications at the various points at which it has been introduced.

It was first introduced in Eqns. (2-3), which give the displacements along the coordinate axes of points originally lying on these axes, in terms of the strain components. We use these equations frequently in this chapter, and at all points the restriction to small strain is introduced. For this reason Eqns. (2-12) giving the displacement of a point not lying on the coordinate axes, and Eqns. (2-19) (and Mohr's circle construction) giving the strains on axes inclined to the original set, are valid only for small strains.

To be strictly correct we should state that Mohr's circle construction and Eqns. (2-19) are valid only when strain is small *if the strain components are defined as in Eqns. (2-3)*. The reason for this distinction is that Mohr's circle and Eqns. (2-19) give the effect of rotation of axes on the components of any second rank symmetric tensor. Large strain is such a quantity and so these equations must be applicable. However, in this case the strain components are defined differently from Eqn. (2-3).

Although the existence of principal axes was derived from small strain equations, the concept is valid at large strains.

The strain ellipsoid illustrates the deformation of a body at large and small strains. However, for large strain the lengths of the principal axes are equal to the principal extension ratios, not to the principal strains added to unity. The definitions of dilatational strain, of pure shear (Section 2-15), and of deviatoric strain are all valid at both large and small strain, but there is an alternative definition of pure shear which is restricted to small strain. Since we used this to analyse deviatoric strain into pure shears, our analysis is invalid at large strain.

The analysis of strain into dilatational and deviatoric components implies that it is possible to superimpose states of strain. Since this is true for small strains only, the analysis is thus restricted. The breakdown of deviatoric strain into pure shears is restricted to small strains for the same reason, in addition to that given earlier. We use the above analyses in Chapter 4 to relate stress and strain, and so these relationships can be used only for small strain.

2-19   Worked examples

1. The principal axes of strain in a sheet of strained material are used as coordinate axes; the principal strains are ε_x and ε_y. A circle of unit radius is drawn with its centre at the origin of coordinates, and the strain is then released. Determine the equation of the figure which is formed.

Consider a point whose coordinates are (x,y) in the unstrained state. On straining, its coordinates become (x',y'), where

$$x' = x(1 + \varepsilon_x) \qquad y' = y(1 + \varepsilon_y)$$

If this point lies on the circle drawn in the strained sheet, then

$$x'^2 + y'^2 = 1$$

Substitution gives

$$x^2(1 + \varepsilon_x)^2 + y^2(1 + \varepsilon_y)^2 = 1$$

and this is the equation of the figure when strain is released. It is the equation of an ellipse whose principal axes coincide with the coordinate axes and which are of length $1/(1 + \varepsilon_x)$ along the x axis and $1/(1 + \varepsilon_y)$ along the y axis.

2. A sheet of material is subjected to large pure shear, the principal extension ratio being α along the x axis and $1/\alpha$ along the y axis. Determine the angle between the x principal axis and a line of particles which are unstrained on deformation

(a) in the strained state, and
(b) in the unstrained state.

The polar equation of the strain ellipsoid is given by Eqn. (2-27), and for large strain with the principal extension ratios given this becomes

$$\frac{\cos^2 \phi}{\alpha^2} + \alpha^2 \sin^2 \phi = \frac{1}{r'^2}$$

A line of particles unextended by the deformation will have unit length and so will be at an angle ϕ, given by

$$\frac{\cos^2 \phi}{\alpha^2} + \alpha^2 \sin^2 \phi = 1$$

Therefore

$$\frac{1}{\alpha^2} + \alpha^2 \tan^2 \phi = \frac{1}{\cos^2 \phi} = 1 + \tan^2 \phi$$

Simplifying gives

$$\tan \phi = 1/\alpha$$

If the angle this line of particles makes with the axis in the unstrained state is θ, then, from Eqn. (2-26),

$$\tan \phi = \frac{1/\alpha}{\alpha} \tan \theta \qquad \text{or} \qquad \tan \theta = \alpha^2 \tan \phi$$

Therefore

$$\tan \theta = \alpha$$

3. A sheet of material is deformed in large strain so that two lines of particles which are mutually perpendicular before straining do not change in length,

but so that each rotates towards the same principal axis through the same angle. Determine the angles between these lines and the principal axes of strain in the unstrained state.

Let OA and OB (Fig. 2-23) be the lines, and let the circle be drawn in the sheet before straining. On straining this will change into the ellipse, but since OA and OB do not change in length their ends will lie on the line traced by

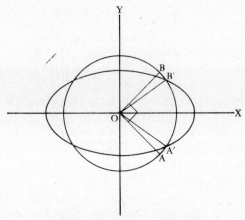

Fig. 2-23   Worked example 2-19(3)

the circle as well as on the ellipse. Thus they must terminate at A' and B', the points of intersection of the circle and the ellipse. Hence, by symmetry

$$\angle A'OX = \angle B'OX$$

where OX is a principal axis. Therefore

$$\angle AOX = \angle BOX$$

But since

$$\angle AOX + \angle BOX = 90°$$

$$\angle AOX = 45°$$

Thus in the unstrained state the lines are at 45° to the principal axes.

Would the same answer be obtained if the lines were allowed to change in length?

Relevant exercises: Nos. 2-11 to 2-14.

2-20   Summary
(i) With respect to any set of cartesian axes, six quantities, three shear and three extensional strain components, are necessary to define strain. Physical quantities such as this are known as second rank symmetric tensors (Sections 2-4 and 2-5).

(ii) One particular set of axes can be found with respect to which the strain can be completely specified by three extensional strains. These axes are called the principal axes of strain, and the three strains are called the principal strains (Section 2-10).

(iii) Small strain can be resolved into dilatational and deviatoric components (Section 2-17).

(a) The dilatational strain is given by

$$\Delta = \varepsilon_x + \varepsilon_y + \varepsilon_z \quad \text{[Eqn. (2-31)]}$$

(b) The component, ε_d, of each principal strain due to the dilatational component is given by

$$\varepsilon_d = \tfrac{1}{3}\Delta \quad \text{[Eqn. (2-29)]}$$

(c) The components of each principal strain due to the deviatoric component are given by Eqns. (2-32).

(d) The deviatoric strain can be broken down into various combinations of pure shears, some of which are illustrated in Fig. 2-21.

(iv) Two principal strains of equal magnitude but of opposite sign, with the third principal strain zero, represent a small pure shear. On axes at 45° to the non-zero principal strains, and in the same plane, the extensional strains are zero and the shear strain is numerically equal to the principal strain (Section 2-15).

EXERCISES

Sections 2-1 to 2-6

2-1 In the following table, five different sets of coordinates are given for the points P and Q. These points lie on the x and y axes respectively, and x and y represent their coordinates. In each case a different state of small strain is applied and x' and y' are the coordinates after straining. Assuming $\varepsilon_{xy} = \varepsilon_{yx}$, calculate the state of strain, and the x coordinate of the point Q.

	P			Q	
x	x'	y'	y	y'	
2·000	2·012	0·004	5·000	4·988	
4·000	4·008	−0·010	3·000	2·996	
6·000	5·980	−0·013	4·000	4·010	
3·000	2·995	0·006	6·000	6·015	
5·000	5·015	0·008	1·000	0·998	

2-2 P, Q, and R represent points lying on the x, y, and z axes respectively, and x, y, and z are their coordinates before deformation. Their coordinates after deformation are x', y', and z'. In the following table, (a), (b), and (c) represent three different states of strain. Fill in the gaps in the table.

		(a)	(b)	(c)
P	x		3·000	5·000
	x'	4·016		5·005
	y'	0·036		
	z'			−0·020
Q	y	2·500	4·000	−3·000
	x'			
	y'	2·505		−3·006
	z'		−0·012	0·006
R	z	8·000	6·000	2·000
	x'		0·012	−0·008
	y'	−0·032		
	z'		5·994	
	ε_{xx}	0·004	−0·005	
	ε_{yy}		0·003	
	ε_{zz}	−0·005		−0·005
	ε_{xy}		−0·004	0·003
	ε_{yz}			
	ε_{xz}	0·0015		

2-3 A square sheet of material has a side of length a. It is deformed by displacing one edge of the square a small distance x in a direction perpendicular to its length, the opposite edge remaining fixed in space. Determine the state of strain as measured by axes drawn along the diagonals of the square.

Sections 2-7 to 2-12

2-4 Coordinate axes are drawn in a sheet of material and a point P is located at perpendicular distances of 2 cm and 3 cm from the x and y axes respectively. The sheet is deformed uniformly and rigid body motions are applied so that the origin of coordinates occupies the same position in space as previously and $\varepsilon_{xy} = \varepsilon_{yx}$. The state of strain is given by $\varepsilon_{xx} = 0.0065$, $\varepsilon_{yy} = -0.0034$, $\varepsilon_{xy} = 0.0055$.

(a) Calculate the components of displacement of P in the x and y directions.

(b) Determine the equation giving the positions of points which will not be displaced in the y direction. Express this equation so that the points can be located: (i) before and (ii) after deformation.

2-5 The strain in a sheet of material with respect to a given set of axes is $\varepsilon_{xx} = 0.003$, $\varepsilon_{yy} = -0.005$, $\varepsilon_{xy} = -0.002$. The axes are rotated in an anticlockwise direction through an angle θ. Plot graphs of ε_{1xx}, ε_{1yy}, and ε_{1xy} against θ and determine from the graphs the directions of the principal axes of strain and the magnitude of the principal strains.

2-6 (a) Devices are available for measuring extensional strain in sheets of material. How many are necessary to determine the state of strain in a lamina and in what directions must they be mounted?

(b) If devices were available for measuring shear strain how many of these would be necessary and in what directions would they have to be mounted?

2-7 A rectangular sheet of material 15 cm by 20 cm is deformed uniformly in small strain so that its edges are of length 15·030 cm and 20·060 cm respectively. The angle between them decreases by 0·010 radian. A hole with straight edges and right-angled corners is cut in the centre of the sheet before deformation. Determine the angle between the edges of the sheet and the edges of the hole, if the corners of the hole remain right angles on deformation. If the edges of the hole are initially 1 cm long, calculate their final length.

2-8 Coordinate axes are drawn in a sheet of material which is then strained. Use Mohr's circle construction to show that the sum of the extensional strains along these axes is independent of their direction.

2-9 In the following problems ε_{xx}, etc., indicate the strain with respect to one set of axes; ε_{1xx}, etc., the strain with respect to axes at angle θ to the first; ε_x, etc., the principal strains; and ϕ the angle between the first set of axes and the principal axes.

(a) Given $\varepsilon_{xy} = 0·003$, $\varepsilon_{1xy} = -0·004$, $\theta = 30°$, determine ϕ.

(b) Given $\varepsilon_{xx} - \varepsilon_{yy} = 0·0005$, $\phi = 30°$, find ε_{xy} and $\varepsilon_x - \varepsilon_y$.

(c) Given $\varepsilon_{xx} - \varepsilon_{yy} = 0·001$, $\varepsilon_{1xx} = -0·002$, $\varepsilon_{1yy} = 0·0008$, $\theta = 60°$, determine ε_{xx}, ε_{yy}, ε_{xy}, ε_{1xy}, ϕ, ε_x, ε_y.

2-10 Given that for a sheet of strained material $\varepsilon_{xx} = 0·008$, $\varepsilon_{yy} = 0·001$, $\varepsilon_{xy} = 0·004$, determine

(a) the principal strains,

(b) the maximum change on straining in the angle between two initially mutually perpendicular lines.

Sections 2-13 to 2-19

2-11 In the following table (x,y,z) represent the coordinates of a point in the undeformed state, and ε_x, ε_y, ε_z the principal strains, which are in the directions of the coordinate axes. In each case determine the change on deformation in the angles (in radians) between the radius vector to the point and the coordinate axes, choosing either positive or negative axes so that this angle is less than $\frac{1}{2}\pi$.

	ε_x	ε_y	ε_z	x	y	z
(a)	0·005	0·001	−0·003	3	−2	0
(b)	0·0015	0·005	0·002	0	1	1
(c)	−0·004	0·008	0·006	−2	0	4

2-12 A block of material is, when unstrained, in the form of a rectangular parallelepiped with edges parallel to the x, y, and z coordinate axes and of lengths 10 cm, 12 cm, and 15 cm respectively. The block is deformed so that the principal axes of strain are (1) along the z axis, (2) rotated through 30° from the positive x direction towards the positive y direction, and (3) at right angles to (1) and (2). The principal strains are $\varepsilon_1 = 5 \times 10^{-3}$, $\varepsilon_2 = -3 \times 10^{-3}$, $\varepsilon_3 = 2 \times 10^{-3}$. Determine
   (a) the change in the angles between the edges,
   (b) the change in the lengths of the edges,
   (c) the change in the volume of the block,
   (d) the deviatoric strain components of the principal strains.
(For (a) and (b) use the equation for the strain ellipse in polar form.)

2-13 A square sheet of material with side of length 10 cm is deformed by sliding one edge in the direction of its length a distance of 0·5 mm, the opposite edge being fixed, and all edges remaining straight and unchanged in length. Determine the principal strains and the directions of the principal axes of strain.

2-14 The principal strains in a deformed body are 1×10^{-3}, 2×10^{-3}, and 3×10^{-3}. If this deformation is to be achieved in two stages, find the principal strains of the intermediate stage in each of the following cases.
   (a) The volume at the intermediate stage is the same as in the undeformed state, and the final size is achieved without changing the angle between any pair of lines drawn in the body.
   (b) The intermediate stage is achieved from the undeformed state without changing the angle between any pair of lines drawn in the body, and the final stage is achieved without change of volume.

3

Specification of Stress

3-1 Contact and body forces

A state of stress is produced in a body when a system of forces acts upon it. These forces can be applied in two ways: they can act directly on every particle of the body, e.g., gravitational and magnetic forces; or they can act directly on the surface and be transmitted indirectly to the interior of the body through its constituent particles, e.g., a heavy weight standing on the body. The first type of force is a *body force*; the second type is a surface or *contact force*. Note that the contact force in the example given was produced by the body force exerted on the weight by gravitation.

In this book we are concerned only with contact forces in static equilibrium. Also, we consider forces distributed uniformly over areas of the body, but for simplicity these are shown in the diagrams as the single, equivalent, resultant force.

3-2 Physical nature of stress

Consider a body acted upon by a system of contact forces in equilibrium [Fig. 3-1(a)]. If the body were cut along the plane shown shaded in the

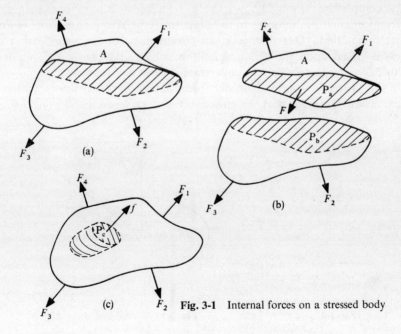

Fig. 3-1 Internal forces on a stressed body

figure, the separate halves would no longer be in equilibrium and would fly apart. However, before cutting, the portion of the body labelled A is in equilibrium, so an *internal force F* must be exerted on the material in the plane P_a by the material in the plane P_b, such that F, F_1, and F_4 form a system of forces in equilibrium. (The two halves are shown separately in Fig. 3-1(b) for clarity.)

In Fig. 3-1 the plane has been drawn with a particular orientation, but the same conclusion would follow whatever orientation was chosen. Thus, if any plane is drawn in a body subjected to a system of forces, the material in this plane will be acted upon by an internal force exerted by the adjacent material. The stress is the quantity which enables this internal force to be determined.

Suppose now that, instead of drawing the plane right through the body, as in Figs. 3-1(a) and 3-1(b), we consider a small plane area P_c [Fig. 3-1(c)], which is part of a closed surface drawn in the interior of the body [shaded in Fig. 3-1(c)]. The material lying in P_c will be subject to a force, f, exerted by the adjacent layer of material lying *outside the shaded surface*. This is the force which would have to be applied to P_c, if the body were cut through along this plane, to hold this portion of the body in equilibrium. We can regard it as a contact force acting on the plane, and distributed over its surface. We define the stress acting on this plane as the magnitude of the force distributed over unit area. If we change the orientation of the plane, the magnitude and direction of the force will be changed. To define the stress completely, we must know its value on a sufficient number of planes of different orientation, in order to calculate the force acting on a plane of any other orientation. Our mathematical development of the specification of stress is based upon the above definition, which also provides a physical picture of considerable help in understanding this development.

In this chapter, we often consider forces acting on planes drawn within a body. These planes are part of a closed surface, and we adopt the conventions

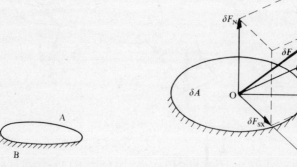

Fig. 3-2   **Fig. 3-3** Normal and shear stresses

that the forces considered as acting on the plane are those exerted by the material lying outside the surface, and that the normals drawn to the planes are directed outwards from the surface. In most diagrams only a small plane area is shown, the remainder of the closed surface being omitted. Where ambiguity exists, the side of the plane lying inside the surface is shaded. For example, for the plane in Fig. 3-2, the side A lies outside the closed surface and the side B lies inside.

3-3 Shear and normal stress

In Fig. 3-3, δA is a small plane area forming part of a closed surface drawn within a stressed body. The remainder of the closed surface lies on the shaded side of the plane. The layer of material adjacent to the plane, lying outside the surface, exerts a force δF on the plane, represented by the vector **OF**. The force δF may be resolved into two components: δF_N, normal to the plane, and δF_s, tangential to the plane. $\delta F_N/\delta A$ is the *normal*, or *tensile*, stress acting on this plane; $\delta F_s/\delta A$ is the *shear* stress acting on the plane. We can draw mutually perpendicular directions, OX and OY, tangential to the plane, and resolve δF_s along these into the two components δF_{sx} and δF_{sy}. So we can represent the stress across this plane by three numbers, one of which is a normal stress, the other two being shear stresses.

3-4 Uniform stress

We chose the area, δA, of the plane layer quite arbitrarily. We could draw different areas in this plane, measure the forces acting on them, and determine the stress. If this stress were constant whatever area we chose, then the stress acting on the plane would be uniform at all points. If we drew a series of different planes, each one parallel to δA, and obtained the same stress for each plane, then the stress in this direction would be uniform throughout the body. If the above conditions were satisfied by any series of parallel planes, whatever their direction, then the body would be in a state of uniform stress.

Apart from one or two special cases which are treated separately, this book is concerned only with states of uniform stress.

3-5 Complete specification of stress

So far we have considered the force acting on one arbitrarily drawn plane and used it to define three stress components. Had we chosen a plane inclined to the first, we would have obtained three different values for these components. Could these values be calculated, knowing the components on the first plane and the inclination between the two planes? In other words, do the stress components on one plane provide a complete specification of the stress, and if not, how much more information is necessary?

Consider a set of cartesian coordinate axes drawn in a stressed body

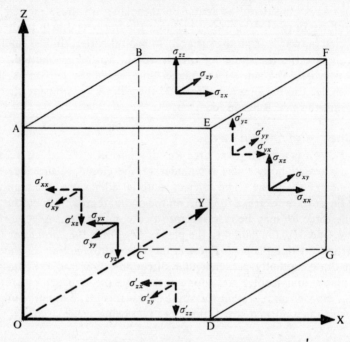

Fig. 3-4 Components of stress

(Fig. 3-4); a small cube is formed by planes normal to them. An internal contact force will be exerted on each face of this cube by the layer of material adjacent to it and lying outside. We can resolve the force acting on a given face into components parallel to the axes, and from these determine the stress components acting on the face (as in Section 3-2). We denote these stress components by the symbol σ with two suffixes. *The first suffix denotes the direction of the axis normal to the face on which the stress acts; the second suffix denotes the axis along which the stress component is directed. If the directions of both these axes have the same sign, the stress component is positive; if they have different signs, it is negative.* The positive stress components acting on each of the faces of the cube are shown in Fig. 3-4.

Three stress components act on each of the six faces of the cube, giving 18 components in all. How many of these are independent? We can answer this by considering conditions of equilibrium and uniform stress. The contact forces acting on the faces of the cube must be in equilibrium. This can best be seen by considering what would happen if they were not. Suppose, for example, that the force on the face DEFG in Fig. 3-4 is greater than that on the opposite face OABC, both forces being directed outward from the cube. The cube will move *with the solid body* in the positive x direction. This will increase the spacing between the molecules in the face OABC and those adjacent to it outside the cube (i.e., the strain in the material will be changed), and so increase the force of attraction between them. Hence the contact force on this face will increase. Similarly, the spacing between the molecules in the face DEFG and their neighbours outside the cube will decrease, causing the contact forces on this face to decrease. This process will continue until the forces are in equilibrium. This argument has said nothing about the externally applied forces, and so must hold whether or not these are in equilibrium. (However, the argument does require the movement of a cube of finite mass and this will generate inertial forces which have been neglected. These are only important when the external forces are applied impulsively.)

Since the stress is uniform, from Section 3-4, equal internal forces act across parallel planes of equal area. Hence, the force acting on the plane DEFG must be equal and opposite to that on OABC. (These forces are oppositely directed because each is the force exerted on the face by the material lying *outside* the cube.) Hence

$$\sigma_{xx} = \sigma'_{xx}$$
$$\sigma_{xy} = \sigma'_{xy}$$
$$\sigma_{xz} = \sigma'_{xz}$$

(The primed quantities denote the stresses acting on the hidden faces in Fig. 3-4.) Similar relationships can be derived for the other faces, reducing the 18 components to nine.

These nine quantities are normally written as an array:

$$\begin{array}{ccc} \sigma_{xx} & \sigma_{xy} & \sigma_{xz} \\ \sigma_{yx} & \sigma_{yy} & \sigma_{yz} \\ \sigma_{zx} & \sigma_{zy} & \sigma_{zz} \end{array}$$

but they are not all independent. We can show this by considering the equilibrium of the cube. Let the length of an edge be l. If we apply the condition of uniformity of stress to the components shown in Fig. 3-4, we see that the cube must be in *translational*† equilibrium along any of the directions of the coordinate axes. So we can derive no further information by considering translational equilibrium. However, consideration of *rotational*† equilibrium is more useful. First of all, take moments about an axis through the centre of the cube, parallel to OY. The clockwise moment is

$$(l^2\sigma_{zx})\tfrac{1}{2}l + (l^2\sigma'_{zx})\tfrac{1}{2}l = \tfrac{1}{2}l^3(\sigma_{zx} + \sigma'_{zx})$$

where $l^2\sigma_{zx}$ is the force parallel to OX acting on face ABFE, and $l^2\sigma'_{zx}$ is similarly defined. The anticlockwise moment is

† A body is in *translational* equilibrium when the forces acting upon it will not produce movement along a straight line; it is in *rotational* equilibrium when there is no resultant couple acting upon it, and hence it has no tendency to rotate.

$$\tfrac{1}{2}l^3(\sigma_{xz} + \sigma'_{xz})$$

By equating these two and using the condition of uniform stress, we get

$$\sigma_{zx} = \sigma_{xz} \tag{3-1a}$$

Similarly, taking moments about axes through the centre, parallel to OX and OZ gives, respectively,

$$\sigma_{zy} = \sigma_{yz} \tag{3-1b}$$

and

$$\sigma_{xy} = \sigma_{yx} \tag{3-1c}$$

We have now reduced the number of stress components to six independent quantities and, from Eqns. (3-1), the array given above must be symmetrical about the principal diagonal.

If we compare this specification with that given for strain in Section 2-5, we see that the two are exactly the same, and so stress must also be a second rank symmetric tensor. In Section 2-5, we saw that such a quantity expressed a relationship between two vectors, strain relating the position vector to the displacement vector. In the same way, stress relates the force vector to the vector denoting area.

Although we derived the results of Chapter 2 by considering strain, many of these results are general properties of any second rank symmetric tensor, and could equally well have been determined by considering any other physical quantity of this type. The equations giving the effect on the strain components of the rotation of axes, the existence of principal axes, and the analysis of strain into dilatational and deviatoric components are all examples of results having such general validity.† We could therefore apply them without proof to the analysis of stress. However, to illustrate the truth of this statement, we derive these results in subsequent sections.

3-6 Worked examples

1. An internal force F acts on a plane of area A drawn in a stressed body, and makes an angle θ to the normal to the plane. Determine the normal and shear forces.

$$\text{Component of force normal to plane} = F \cos \theta$$

$$\text{Therefore, normal stress} = \frac{F \cos \theta}{A}$$

$$\text{Component of force tangential to plane} = F \sin \theta$$

$$\text{Therefore, shear stress} = \frac{F \sin \theta}{A}$$

† The parallelogram of forces can be quoted as an example, already familiar to the student, of a result first proved for a specific physical quantity, but having a general validity. It is usually first considered as a means of adding forces, but it is a construction which can be used to add any vector quantity.

2. Coordinate axes are taken as the edges of a rectangular parallelepiped of material, the lengths of the edges along the x, y, z axes being 5 cm, 7 cm, and 4 cm, respectively. An equilibrium set of outward acting forces is distributed uniformly over the surfaces. The magnitude of the resultant force on each face and the cosines of the angles between its line of action and the coordinate axes are given in the following table. Write down the array of numbers specifying the state of stress.

Axis normal to face	Magnitude of force (dyn)	Cosine of angle between force and	
		x axis	y axis
OX	9	0·8	0·5
OY	11		0·4
OZ	20		

Consider the face normal to the x axis. Area $= 7 \times 4 = 28$ cm². Component of force acting on it parallel to the x axis is

$$9 \times 0\cdot8 = 7\cdot2 \text{ dyn}$$

Therefore $\qquad\qquad \sigma_{xx} = 7\cdot2/28 = 0\cdot258 \text{ dyn/cm}^2$

Component parallel to y axis is

$$9 \times 0\cdot5 = 4\cdot5 \text{ dyn}$$

Therefore $\qquad \sigma_{xy} = 4\cdot5/28 = 0\cdot161 \text{ dyn/cm}^2 = \sigma_{yx}$

Since the sum of the squares of the direction cosines of a line are equal to unity, then if the direction cosine of the line to the z axis is α_z,

$$\alpha_z^2 = 1 - (0\cdot8^2 + 0\cdot5^2) = 0\cdot11$$

giving $\qquad\qquad\qquad \alpha_z = \pm0\cdot332$

Therefore the component of force parallel to the z axis is

$$\pm9 \times 0\cdot332 = \pm2\cdot99 \text{ dyn}$$

and so

$$\sigma_{xz} = \pm2\cdot99/28 = \pm0\cdot107 \text{ dyn/cm}^2 = \sigma_{zx}$$

Consider, next, the face normal to the y axis. Area $= 20$ cm². Component of force acting on it parallel to y axis is

$$11 \times 0\cdot4 = 4\cdot4 \text{ dyn}$$

Therefore $\qquad\qquad \sigma_{yy} = 0\cdot22 \text{ dyn/cm}^2$

To determine σ_{yz} we need to know the direction cosine between the force on this face and the z axis, which is not given. However, it can be determined from the direction cosine to the x axis (which can be calculated) and the fact that the sum of the squares of the three direction cosines is unity. Let the direction cosine of the force to the x axis be β_x and to the z axis be β_z. Then, component of force parallel to x axis $= 11\beta_x$ and $\sigma_{yx} = 11\beta_x/20$. We have already determined $\sigma_{yx} = 0.161$, giving $\beta_x = 0.293$. Since

$$\beta_x^2 + 0.4^2 + \beta_z^2 = 1$$

$$\beta_z = \pm 0.869$$

and hence $\sigma_{yz} = \pm 0.478 = \sigma_{zy}$

Now consider the face normal to the z axis which has an area of 35 cm². Let the direction cosines of the force acting on this face to the x, y, z axes be γ_x, γ_y, γ_z, respectively. We can find γ_x from σ_{zx} and γ_y from σ_{zy} in the same way as β_x was found from σ_{yx} above, giving $\gamma_x = \pm 0.187$ and $\gamma_y = \pm 0.835$. Then, since

$$\gamma_x^2 + \gamma_y^2 + \gamma_z^2 = 1$$

$$\gamma_z = 0.517$$

(Since it is given that the force is outward acting, γ_z must be positive.) We can now determine σ_{zz}, which is 0.296 dyn/cm². Hence the state of stress is given by the array

$$
\begin{array}{ccc}
0.258 & 0.161 & \pm 0.107 \\
0.161 & 0.22 & \pm 0.478 \\
\pm 0.107 & \pm 0.478 & 0.296
\end{array}
$$

We are not given sufficient information to determine the sign of some of the shear stresses.

Relevant exercises: Nos. 3-1 to 3-2.

3-7   Effect of rotation of axes

So far, we have seen that, if a set of cartesian axes is drawn in a block of material, the stress may be specified by the forces on small planes normal to these axes, and that in order to specify the stress completely in this way, we need six components. Knowing this, it should be possible, therefore, to determine the stress components for any other set of coordinate axes inclined to the first set. To do this, we consider the case in which the axis of rotation coincides with one of the coordinate axes, and for which the lines of action of the forces all lie in the plane normal to the axis of rotation. This is much simpler than the general three-dimensional case.

We denote the original axes by OX, OY, and OZ, and planes OADC, OAB, and OCEB are drawn perpendicular to these directions (Fig. 3-5). OX_1, lying in the XZ plane, denotes the x direction of the new set of axes, and the plane ADEB is drawn normal to this direction forming the wedge of material shown in Fig. 3-5. The stress components acting on the faces OADC and OCEB of this wedge are σ_{xx}, σ_{zz}, and σ_{xz}; those on the face OAB must be zero to satisfy the condition that all stress components lie in the XZ plane. The stress components on the face ADEB are σ_{1xx} and σ_{1xz}, and we have to express these in terms of σ_{xx}, σ_{xz}, σ_{zz}, and the angle θ between the old and new x directions.

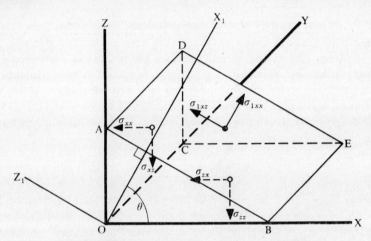

Fig. 3-5   Effect of rotation of axes on stress components

The block of material must be in equilibrium under the internal forces acting on its faces. However, we cannot immediately write down the equations of equilibrium in terms of the stresses. Since stress is a second rank tensor quantity the stress components *cannot* be resolved along coordinate axes by the rules of vector algebra. They must first of all be converted to forces, which *are* vector quantities. We do this by multiplying each stress component by the area of the face on which it acts. If the area of the plane ABED is A, then the area of the plane OCEB is $A \sin \theta$, and that of the plane OCDA is $A \cos \theta$, enabling the forces acting on the planes to be determined. Resolving along the direction OX gives, for equilibrium,

$$-\sigma_{zx}A \sin \theta - \sigma_{xx}A \cos \theta + \sigma_{1xx}A \cos \theta - \sigma_{1xz}A \sin \theta = 0$$

Resolving along OZ

$$\sigma_{1xx}A \sin \theta + \sigma_{1xz}A \cos \theta - \sigma_{zz}A \sin \theta - \sigma_{xz}A \cos \theta = 0$$

Solving these equations for σ_{1xx} and σ_{1xz}

$$\sigma_{1xx} = \sigma_{xx} \cos^2 \theta + \sigma_{zz} \sin^2 \theta + \sigma_{xz} \sin 2\theta \qquad (3\text{-}2a)$$

$$\sigma_{1xz} = \tfrac{1}{2}(\sigma_{zz} - \sigma_{xx}) \sin 2\theta + \sigma_{zx} \cos 2\theta \qquad (3\text{-}2b)$$

If we substitute $(\theta + \frac{1}{2}\pi)$ for θ in these equations, then the left-hand side of Eqn. (3-2a) becomes σ_{1zz} and the left-hand side of Eqn. (3-2b) becomes $-\sigma_{1zx}$ (this is negative because it acts in the negative x direction). Whence

$$\sigma_{1zz} = \sigma_{xx} \sin^2 \theta + \sigma_{zz} \cos^2 \theta - \sigma_{xz} \sin 2\theta \tag{3-2c}$$

and $\qquad \sigma_{1xz} = \sigma_{1zx}$

3-8 Principal axes of stress and principal stresses

Equations (3-2) show the way in which the numbers used to define a given state of stress vary as the coordinate axes are rotated in a stressed body. They are identical in form with Eqns. (2-19) which give the equivalent result for strain. Hence, the conclusions which followed from Eqns. (2-19) for a strain must also apply to stress. In particular, if we consider Eqn. (3-2b), which describes the variation of shear stress, then, on axes at an angle θ to the original, given by

$$\tan 2\theta = \frac{2\sigma_{xz}}{\sigma_{xx} - \sigma_{zz}} \tag{3-3}$$

the shear stress must be zero. As already pointed out in Section 2-10, there is always one value of θ between 0 and $\frac{1}{2}\pi$ which satisfies this equation, whatever the values of σ_{xx}, σ_{zz}, and σ_{xz}. (There are in fact two solutions differing by $\frac{1}{2}\pi$ between 0 and π. These represent the x and z directions for which the shear stress is zero.)

So we see that, providing all the stress components lie in one plane which contains two of the coordinate directions, then another pair of coordinate directions exist in this plane for which the shear stresses are zero. It is possible to prove an equivalent result for the general three-dimensional case, and we can say that, *whatever the state of stress, there always exists one set of three mutually perpendicular directions for which the shear stress components are zero.*

These directions are the *principal axes of stress*. With respect to this particular set of axes, and to this set alone, the state of stress in a three-dimensional body can be completely specified by three normal stresses acting along the principal axes. These are the *principal stresses*. In this book we denote them by the use of a single suffix—the same convention that we used for strain. Thus σ_x is the normal stress acting on a plane normal to the x principal axis.

As for strain, problems are often greatly simplified if the principal axes of stress can be located. When all the stress components lie in one plane, or if one of the principal directions is known, we can locate the axes using Eqn. (3-3). Mohr's circle construction (Section 2-11) can also be used. This was shown to be a valid construction from Eqns. (2-19), and since Eqns. (3-2) are identical with these, it can also be used to locate principal axes of stress, determine principal stresses, and investigate the variation of stress components as the coordinate axes are rotated.

From this construction, we can derive results which are analogous to those derived for strain. These results are:

(a) The maximum and minimum values of σ_{xx} and σ_{zz}, obtained as the coordinate axes are rotated, are those of the principal stresses.

(b) The maximum value of the shear stress is obtained when the axes are at 45° to the principal axes.

(c) On axes at such an angle the normal stresses are equal.

(d) If the principal stresses are of equal magnitude but opposite sign, then on axes at 45° to these the normal stresses are zero, and the shear stress is equal to the principal stress.

This is referred to as a *pure shear* stress.

3-9 Worked examples

1. A cube of metal with side of length l is immersed in a fluid at a pressure P. A tensile force F is applied to one pair of opposite faces and acts normally to these faces. Determine (a) the magnitudes and directions of the principal stresses, and (b) the stress components acting on a diagonal plane whose normal makes an angle of 45° to the tensile forces.

(a) Coordinate axes are chosen as in Fig. 3-6(a), and the forces acting on

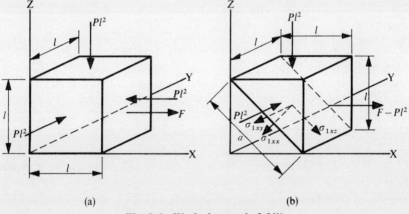

(a)   (b)

Fig. 3-6 Worked example 3-9(1)

the faces of the cube are as illustrated. (Forces on the hidden faces have been omitted for clarity.) We can see from this figure that the three resultant forces are mutually perpendicular and normal to the faces of the cube. Hence, there are no shear components with the coordinate axes chosen, which must therefore be principal axes.

The principal stresses are

$$\sigma_x = \frac{F - Pl^2}{l^2} = \frac{F}{l^2} - P$$

$$\sigma_y = \sigma_z = -P$$

(Outward acting stresses are positive.)

(b) Let the stress components acting on the diagonal plane be σ_{1xx}, σ_{1xy}, σ_{1xz}, as in Fig. 3-6(b). Then, considering the equilibrium of the section shown, resolving along OX,

$$F - Pl^2 + a l\sigma_{1xz} \cos 45° - a l\sigma_{1xx} \cos 45° = 0$$

resolving along OY

$$\frac{Pl^2}{2} - \frac{Pl^2}{2} - a l\sigma_{1xy} = 0$$

resolving along OZ

$$Pl^2 + a l\sigma_{1xx} \cos 45° + a l\sigma_{1xz} \cos 45° = 0$$

Now $a = 2^{1/2}l$; $\cos 45° = 2^{-1/2}$, and substituting in the above equations gives

$$\sigma_{1xz} - \sigma_{1xx} = -\frac{F - Pl^2}{l^2}$$

$$\sigma_{1xy} = 0$$

$$\sigma_{1xz} + \sigma_{1xx} = -P$$

whence $\sigma_{1xz} = -F/2l^2$ $\sigma_{1xx} = F/2l^2 - P$

2. In Fig. 3-7(a) the diagram represents a section through a cube of material, the forces being applied so as to produce a uniform distribution of stress. (The broken arrows show the distribution of the force of 7 dyn over the faces to which it is applied.) Determine the magnitude and direction of the principal stresses.

If the force of 7 dyn is applied so as to produce a uniform distribution of stress, the normal stress on the plane AB due to this force must equal that on CD.

AB $= 2 \sin 30°$, hence area of plane AB $= 4 \sin 30° = 2$ cm²

CD $= 2 \cos 30°$, hence area of plane CD $= 4 \cos 30° = 3{\cdot}46$ cm²

Let the force acting across the plane AB be F_1; then the force acting across the plane CD is $7 - F_1$. Therefore $\frac{1}{2}F_1 = (7 - F_1)/3{\cdot}46$, whence $F_1 = 2{\cdot}57$ dyn. Hence the forces can be represented as in Fig. 3-7(b).

If coordinate axes are taken normal to the faces of the cube, these forces can be resolved along the x and y directions, and their components are as

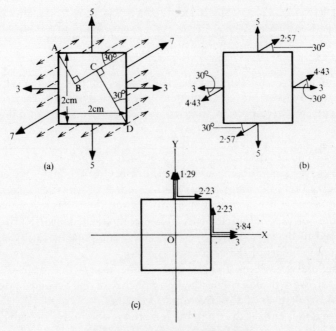

Fig. 3-7 Worked example 3-9(2)

shown in Fig. 3-7(c). The stress components with respect to these axes are therefore

$$\sigma_{xx} = 6.84/4 = 1.71 \text{ dyn/cm}^2$$

$$\sigma_{yy} = 6.29/4 = 1.57 \text{ dyn/cm}^2$$

$$\sigma_{xy} = \sigma_{yx} = 2.23/4 = 0.556 \text{ dyn/cm}^2$$

The principal axes are at an angle θ anticlockwise to the coordinate axes, given by Eqn. (3-3). Substituting the values above in this equation gives

$$\tan 2\theta = \frac{1.112}{1.71 - 1.57}$$

whence

$$\theta = 41°22'$$

Then, substituting in Eqns. (3-2),

$$\sigma_{1x} = 2.20 \text{ dyn/cm}^2, \qquad \sigma_{1y} = 1.08 \text{ dyn/cm}^2$$

σ_{1x} and σ_{1y} are the principal stresses, σ_{1x} acting along the axis $41°22'$ anticlockwise to the x axis. In this problem, the *forces* acting were given, and so could be resolved directly. Had *stresses* been given they would have had to be converted to forces before they could be resolved to determine the components acting along given coordinate directions. This is illustrated in the next example.

3. A body is acted upon by a two-dimensional stress system which, on a given set of axes, has the components $\sigma_{xx} = 7$ dyn/cm², $\sigma_{zz} = 5$ dyn/cm², $\sigma_{xz} = 3$ dyn/cm². A second two-dimensional stress system is now superimposed. The stress components of this system, on a set of axes rotated through an angle of 30° anticlockwise to the first, but lying in the same plane, are $\sigma_{1xx} = 4$ dyn/cm², $\sigma_{1zz} = 8$ dyn/cm², $\sigma_{1xz} = 0$ dyn/cm². Determine the components, with respect to the first set of axes, of the combined stress system.

In this problem, stress components are given with respect to two different sets of axes, the second being rotated with respect to the first. We could use Eqns. (3-2) to find the components of the second system with respect to the first set of axes. Then, since an element of a resultant tensor is equal to the sum of the corresponding elements of the component tensors, provided all are referred to the same set of axes, the stress components can be added to give the components of the combined system.

The method we will use follows this scheme, except that instead of using Eqns. (3-2), we will determine the components of the second system with respect to the first set of axes from first principles.

Suppose a unit cube is drawn in the material of the body, the edges of the cube being parallel to the first set of axes [Fig. 3-8(a)]. Consider now the

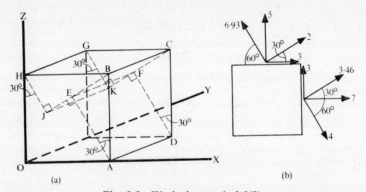

Fig. 3-8 Worked example 3-9(3)

triangular prism of material ABCDEF and let the stress component σ_{1xx} be applied. (For the time being, we will only consider this component; we will consider the effects of the others later.) A force normal to the plane ADFE will then be exerted on the material in this plane by the adjacent material lying just outside the prism; but the force acting on the material in the plane BEFC will be zero. The area of the face ADFE is equal to cos 30°, and so the force acting on it is σ_{1xx} cos 30°, or, since σ_{1xx} is equal to 4 dyn/cm², 3·46 dyn. For the prism to be in equilibrium an equal and opposite force must act on the plane ABCD and so, due to the stress component σ_{1xx}, there is a force of magnitude 3·46 dyn acting on the face normal to the x axis at an angle of 30° to this axis. This is shown in Fig. 3-8(b), which shows a section through the cube parallel to the XZ plane.

For the prism BCGHJK [Fig. 3-8(a)], we can show similarly that σ_{1xx} causes a normal force of 2 dyn to act on the plane GHJK of this prism, and so we may deduce that a force of 2 dyn acts on the face normal to the z axis in the direction shown in Fig. 3-8(b).

Now suppose the stress component σ_{1zz} is applied, and, for the time being, all other stress components are zero. This will cause normal forces to act on the faces BCFE and BCKJ of the prisms ABCDEF and BCGHJK, respectively [Fig. 3-8(a)]. Following the same procedure as above, we can calculate the magnitude of these forces as 4 dyn and 6·93 dyn, respectively. Thus, forces of these magnitudes must act on the faces normal to the x and z axes [Fig. 3-8(b)].

The stress component σ_{1xz} is equal to zero, and so there will be no forces acting due to this component.

Consider now the forces due to all stress components of both systems acting together. These forces are all included in Fig. 3-8(b), those due to the first stress system being the forces normal and tangential to the cube faces.

On the face normal to the x axis there will be a normal force of

$$7 + 3·46 \cos 30° + 4 \cos 60° = 12 \text{ dyn}$$

and a tangential force of

$$3 + 3·46 \cos 60° - 4 \cos 30° = 1·27 \text{ dyn}$$

Since the face is of unit area, if suffix c denotes the stress components due to the combined system,

$$\sigma_{cxx} = 12 \text{ dyn/cm}^2$$
$$\sigma_{cxz} = 1·27 \text{ dyn/cm}^2$$

Similarly

$$\sigma_{czz} = 5 + 6·93 \cos 30° + 2 \cos 60° = 12 \text{ dyn/cm}^2$$

Relevant exercises: Nos. 3-3 to 3-5.

3-10   The stress ellipsoid

We saw in Section 2-13 that a unit sphere drawn in a block of material becomes an ellipsoid, called the strain ellipsoid, when the body is deformed. Because the mathematical properties of stress and strain are identical, we would expect a similar result for stress. The *stress ellipsoid* is defined as follows.

A plane of unit area (Fig. 3-9) is drawn in a block of material under stress. A point O is marked on this plane and a vector **OF** is drawn to represent the total contact force F exerted on the plane by the layer of material adjacent to it on the side of the outward drawn normal. As the orientation of the plane is changed, keeping the point O fixed in space, the surface traced by the point F will be an ellipsoid, which is called the stress ellipsoid. We shall now prove this for two dimensions.

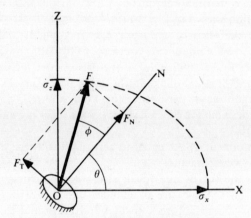

Fig. 3-9 Stress ellipsoid

Let the principal axes of stress be OX and OZ (Fig. 3-9) and let the principal stresses be σ_x and σ_z. Let the normal ON to the plane lie in the XZ plane and make an angle ϕ with **OF**.

The force F can be resolved into components F_n and F_t, normal and tangential to the plane. Remembering that the plane is of unit area, and that σ_x and σ_z are principal stresses so that σ_{xz} is zero, we can write F_n and F_t from Eqns. (3-2)

$$F_n = \sigma_x \cos^2 \theta + \sigma_z \sin^2 \theta \tag{3-4a}$$

$$F_t = (\sigma_z - \sigma_x) \sin \theta \cos \theta \tag{3-4b}$$

Also, from Fig. 3-9,

$$F_n = F \cos \phi \tag{3-5a}$$

$$F_t = F \sin \phi \tag{3-5b}$$

Substituting from Eqn. (3-5a) in (3-4a) and multiplying through by $\sin \theta$ gives

$$F \sin \theta \cos \phi = \sigma_x \sin \theta \cos^2 \theta + \sigma_z \sin^3 \theta \tag{3-6a}$$

Similarly, from Eqns. (3-4b) and (3-5b),

$$F \cos \theta \sin \phi = \sigma_z \sin \theta \cos^2 \theta - \sigma_x \sin \theta \cos^2 \theta \tag{3-6b}$$

Adding together Eqns. (3-6a) and (3-6b) gives

$$F \sin (\theta + \phi) = \sigma_z \sin \theta \tag{3-7a}$$

By a similar procedure we can derive the equation

$$F \cos (\theta + \phi) = \sigma_x \cos \theta \tag{3-7b}$$

We can eliminate θ from the right-hand side of Eqns. (3-7a) and (3-7b) by squaring and adding:

$$\frac{F^2 \sin^2 (\theta + \phi)}{\sigma_z^2} + \frac{F^2 \cos^2 (\theta + \phi)}{\sigma_x^2} = 1 \tag{3-8}$$

This is the polar equation of an ellipse whose major and minor axes lie along OX and OZ and whose radius vector **OF** makes an angle $\theta + \phi$ with OX.

Thus, as the orientation of the unit plane drawn in the material is changed, the end of the vector **OF** will trace out an ellipse. Furthermore, the major and minor axes of this ellipse will coincide with the principal axes of stress, and their lengths will be proportional to the principal stresses (with the same constant of proportionality as that relating the length of **OF** to the force on the plane). An equivalent result holds for three dimensions. However, the stress ellipsoid does not give the convenient picture of the state of stress that the strain ellipsoid does for the state of strain.

One other important equation follows from the discussion above. Dividing Eqn. (3-7a) by (3-7b) gives

$$\tan (\theta + \phi) = \frac{\sigma_z}{\sigma_x} \tan \theta \tag{3-9}$$

This equation enables us to derive the inclination of the force F for a given inclination of the plane, knowing the principal stresses.

3-11 Dilatational stress

If the principal stresses are equal, then in Eqns. (3-4) we can make the substitution $\sigma_x = \sigma_z = \sigma$. This gives

$$F_n = \sigma$$
$$F_t = 0$$

Since F_t is zero, it follows that the force acting on a plane drawn in the material is always normal to the plane, whatever its inclination within the material. It also follows from the first of these equations that the force acting on a unit plane is constant, independent of the inclination of the plane.

We can state this conclusion for the three-dimensional case as follows: if the three principal stresses are equal, then on any plane drawn in the material, whatever its inclination, the shear stress components are zero and the normal stress component is equal to the principal stress. This is an analogous conclusion to that in Section 2-14 for the case of equal principal strains. Hence it is referred to as *dilatational stress*. Since this state can be realized experimentally by immersing a body in a fluid under pressure, it is often called hydrostatic pressure, the pressure in the fluid being equal to the principal stress.

An equivalent result to that proved above can also be obtained by substituting $\sigma_x = \sigma_z = \sigma$ in Eqns. (3-8) and (3-9). Substituting in Eqn. (3-9) gives $\theta + \phi = \theta$, whence $\phi = 0$, which proves that the force on a unit plane drawn in the material is always normal to that plane.

Substituting in Eqn. (3-8) gives $F = \sigma$, proving that the stress ellipsoid is a sphere. These results are shown for two dimensions in Fig. 3-10. Various

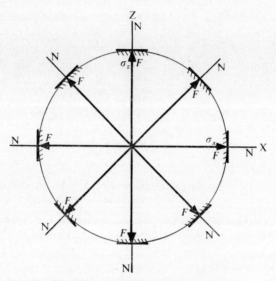

Fig. 3-10 Stress ellipse in dilatation

directions of the force F are shown, and at the point where these vectors touch the stress ellipse a line is drawn to represent the orientation of the plane on which they act. The figure is drawn in this way to avoid confusion; actually, the force F would be exerted on a plane at the centre of the sphere. The force is exerted by the layer of material adjacent to the plane on the unshaded side. So each plane in Fig. 3-10 should be imagined as being at the centre of the sphere and the force F applied to the unshaded side.

The volume of the *strain* ellipsoid has a definite physical significance; it represents the resultant volume under strain of that material lying originally within a sphere of unit radius. This enables volume changes to be distinguished from shape changes and the dilatational strain, Δ, to be defined in terms of volume changes. The volume of the *stress* ellipsoid, however, does not have such simple significance and, in particular, it cannot be used to define dilatational stress.

The only quantity of importance in this case is the principal stress in a dilatational system, which is denoted by σ_d.

Note also that, as for strain, there are no unique principal axes. Any three mutually perpendicular directions can be chosen.

3-12   Stress ellipsoid for pure shear

In Section 3-8, we define a pure shear stress for a two-dimensional state as one for which the principal stresses are of equal magnitude, but of opposite sign. To extend this definition to three dimensions, we need only state that the third principal stress is zero. For such a state of stress, the normal stress components are zero on axes at 45° to the non-zero principal stresses, and in the same plane, and the shear stress components are equal in magnitude to the principal stresses (Section 3-8). The stress ellipse in the plane of the non-zero principal stresses can be derived by making the substitution

$$\sigma_x = -\sigma_z = \sigma$$

in Eqn. (3-8). This gives $F = \sigma$

and so the stress ellipse is a circle.

Now for dilatational stress, in any plane the stress ellipse is also a circle, so the shape of the *stress* ellipsoid, unlike that of the *strain* ellipsoid, does not provide a complete picture of the state of stress. The angle between the normal to the unit plane and the force acting on this plane must also be known (Fig. 3-11).

In Fig. 3-11 the direction of the force F is obtained by considering the equilibrium of a wedge of material formed by the unit plane and the planes normal to the principal axes. The force F is the internal force exerted on the unit plane by the layer of material adjacent to it, and lying outside the wedge, that is, in the direction of the normal N. Similarly, the internal forces exerted on the faces of the wedge normal to the principal axes are in the direction of the principal stresses and can be determined from them. These are represented by the arrows σ_x and σ_z in Fig. 3-11, which shows the development of the ellipse as the plane is rotated. Another way of illustrating the importance of the angle between the normal to the plane and the force acting on it in interpreting the stress ellipse is shown in Fig. 3-12. Different directions of F are taken and, as in Fig. 3-10, a line is drawn where the vector representing F touches the stress ellipse, to represent the orientation of the plane on which the force acts. To interpret this diagram, imagine the plane to be transposed, without altering its orientation, to the origin of coordinates. It is only drawn in the position shown to avoid overcrowding the diagram. The force F is actually exerted on the plane by the layer of material adjacent to it on the unshaded side. If we compare Figs. 3-10 and 3-12, we see that,

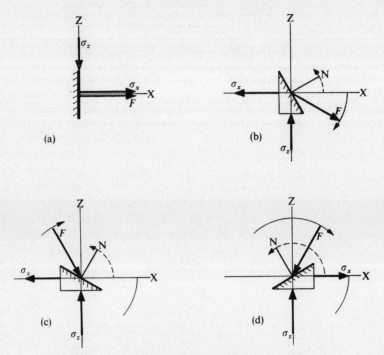

Fig. 3-11 (above and right) Stress ellipse for pure shear

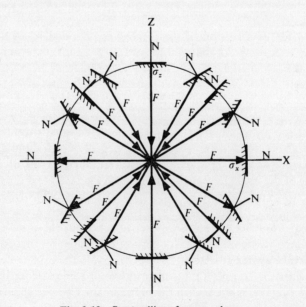

Fig. 3-12 Stress ellipse for pure shear

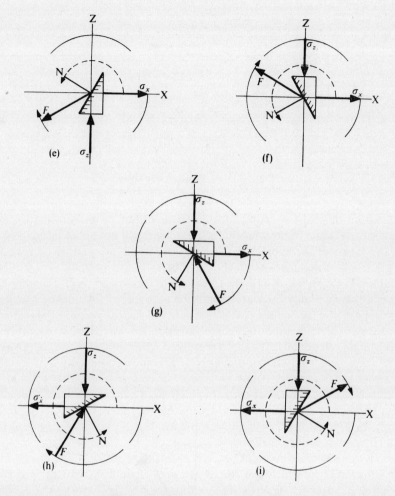

(e) (f)

(g)

(h) (i)

whereas the shape of the stress ellipse is the same in both cases, the inclinations of the planes are very different.

In Section 2-15 it is shown that the volume of a piece of material is unchanged during deformation in pure shear. This follows from the fact that the volume of the *strain* ellipsoid after such deformation is the same as that of a sphere of unit radius. From the discussion above, we see that it is not possible to form any equivalent conclusion for the area of the *stress* ellipse in pure shear.

3-13 Simple shear

If a small cube is drawn in a block of material subject to pure shear, the faces of the cube being normal to the principal axes of stress, then the stress components on these faces will be as shown in Fig. 3-13(a). Now suppose

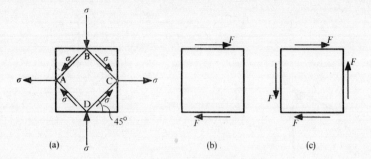

Fig. 3-13   Pure and simple shear

that within this cube a second cube ABCD is drawn. From the analysis of a pure shear stress, the stress components shown will act on its faces.

Now consider the cube in Fig. 3-13(b). This cube can either be part of a larger block of material, in which case the forces are internal, or it can be a complete body, in which case the forces are externally applied. In either case there will be a resultant couple acting on it giving it a tendency to rotate. If it is part of a larger body, this rotation will distort surrounding material until an equal and opposite couple is generated to maintain equilibrium. If it is a complete body, it can be prevented from spinning by clamping one of its faces in a fixed direction in space. The couple applied will then cause distortion with rotation until an opposing couple, sufficient to maintain equilibrium, is generated at the clamp. Thus, in either case, since at equilibrium $\sigma_{xy} = \sigma_{yx}$, the body will eventually be subjected to the set of forces shown in Fig. 3-13(c). By comparison with Fig. 3-13(a) it will be seen that this constitutes a pure shear stress.

Now suppose the constraint which prevents the body spinning is such as to keep the faces to which the forces are applied in Fig. 3-13(b) in a fixed direction in space (these faces are at 45° to the principal axes). The arrangement is then known as *simple shear*. It differs from pure shear in that, for pure shear the *forces applied* are in rotational equilibrium, whereas for simple shear *the direction of the planes to which the shear forces are applied are fixed in space, and equilibrium is only achieved after rigid body rotation*† has distorted surrounding material or generated reactions at clamps. Thus we would expect a simple shear stress to produce the same state of strain as a pure shear stress, but the former will also cause a rigid body rotation.

† Since the direction of one set of parallel planes is fixed, it might be thought that the body would not rotate rigidly. However, since the body is deformable, the directions of other planes inclined to these will be able to change, and this will cause an average rotation of the whole body. An example of this is shown in Fig. 2-17, where the direction of planes parallel to OX remains fixed, but that of planes parallel to OY changes, causing an average rotation of the whole body in a clockwise direction.

3-14 Deviatoric stress

In Section 2-16 we define a deviatoric *strain* as one which changes the shape of a body but not its volume. Because volume changes in the body could be determined quite simply from volume changes of the strain ellipsoid, this provides a convenient definition for the subsequent analysis of a strain into its dilatational and deviatoric components. It is shown above that volume changes in the *stress* ellipsoid do not have such a simple physical interpretation. So we cannot define a *deviatoric stress* in terms of the volume of the stress ellipsoid.

However, we can obtain a definition by analogy with deviatoric strain. In Section 2-17, it is shown that the dilatational strain is equal to the sum of the principal strains [Eqn. (2-31)]. Since, by definition, the dilatation is zero in deviatoric strain, it follows that, for such a strain, the sum of the principal strains is zero [Eqn. (2-33)]. Hence, by analogy, we can define *deviatoric stress as that state of stress for which the sum of the three principal stresses is zero*. Using suffix d to denote a deviatoric stress, then

$$\sigma_{xd} + \sigma_{yd} + \sigma_{zd} = 0 \qquad (3\text{-}10)$$

We also saw in Section 2-17 that a deviatoric strain could be regarded as a superposition of small pure shear strains in the principal planes. Since pure shears in stress and small strain are similarly defined (Sections 2-11 and 3-8), it follows that a deviatoric stress can be regarded as a superposition of pure shear stresses in one or more principal planes (Fig. 3-14). Adding together the three pure shears in Fig. 3-14 gives

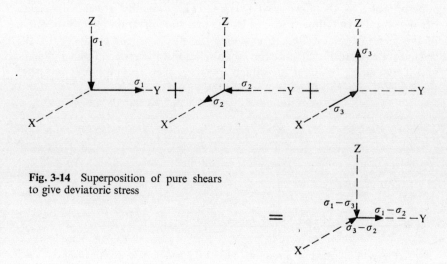

Fig. 3-14 Superposition of pure shears to give deviatoric stress

$$\sigma_{xd} = \sigma_2 - \sigma_3$$

$$\sigma_{yd} = \sigma_1 - \sigma_2$$

$$\sigma_{zd} = \sigma_3 - \sigma_1$$

Adding these together gives Eqn. (3-10), the definition of deviatoric stress.

Although a deviatoric stress is comprised of a superposition of pure shears in the principal planes, if an analysis into these constituent shears were tried, it would be found that, as with strain, no unique solution was possible.

3-15　Analysis of stress into dilatational and deviatoric components

A general state of *strain* can be analysed into its deviatoric and dilatational components by using the definition of these strains in terms of volume changes, and the relationship between the change of volume of the material on straining and the volume of the strain ellipsoid. We cannot use this method for analysing *stress*, since we have seen that the volume of the stress ellipsoid does not have a simple physical interpretation; we have to use an algebraic method. We need to find the dilatational stress, represented as three equal principal stresses σ_d, which, when subtracted from the general state of stress, represented by the three principal stresses σ_x, σ_y, σ_z, will leave a state of stress whose principal values satisfy Eqn. (3-10).

We must make the subtraction according to the rules of tensor algebra (Section 2-17). However, since we can choose the principal axes of the dilatational stress to coincide with those of the general state of stress being analysed, this presents no difficulty, σ_d is simply subtracted from σ_x, σ_y, and σ_z. Also, there is no restriction to small stress for the state of stress to be superposable, as there was with strain. The difficulty arose with strain because the deformation produced by the first state of strain altered the value of the position vector used to calculate the displacement produced by the second state of strain. Stress expresses a relationship between force and area, and if two states of stress are added together, the first state does not alter the value of either of these vectors for the second.

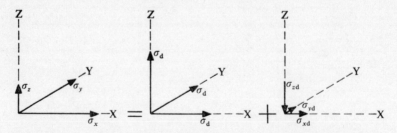

Fig. 3-15　Analysis of stress into dilatational and deviatoric components

Thus the deviatoric components of the principal stresses are given by

$$\sigma_{xd} = \sigma_x - \sigma_d \tag{3-11a}$$

$$\sigma_{yd} = \sigma_y - \sigma_d \tag{3-11b}$$

$$\sigma_{zd} = \sigma_z - \sigma_d \tag{3-11c}$$

Since
$$\sigma_{xd} + \sigma_{yd} + \sigma_{zd} = 0$$

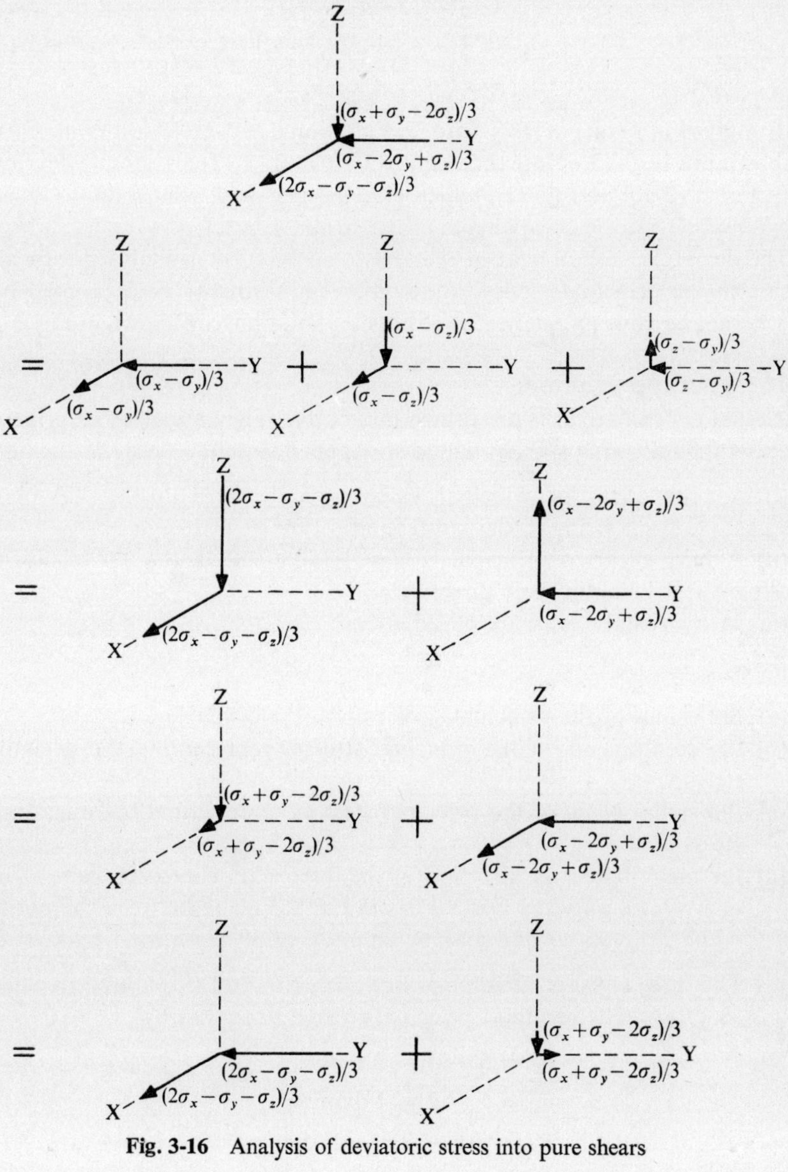

Fig. 3-16 Analysis of deviatoric stress into pure shears

adding together Eqns. (3-11a), (3-11b), and (3-11c) gives

$$\sigma_d = \tfrac{1}{3}(\sigma_x + \sigma_y + \sigma_z) \tag{3-12}$$

That is to say, the dilatational stress component is equal to the average of the principal stresses.

Substituting from Eqn. (3-12) into (3-11) gives

$$\sigma_{xd} = \tfrac{1}{3}(2\sigma_x - \sigma_y - \sigma_z) \tag{3-13a}$$

$$\sigma_{yd} = \tfrac{1}{3}(-\sigma_x + 2\sigma_y - \sigma_z) \tag{3-13b}$$

$$\sigma_{zd} = \tfrac{1}{3}(-\sigma_x - \sigma_y + 2\sigma_z) \tag{3-13c}$$

This analysis of a state of stress into its dilatational and deviatoric components is shown in Fig. 3-15. Note that Eqns. (3-12) and (3-13) are identical in form with Eqns. (2-30) and (2-32), which give the principal values of the dilatational and deviatoric components of the strain.

The deviatoric component of the stress can be further analysed into pure shears in the principal planes, although, as has already been pointed out, this cannot be done uniquely. Some possible solutions are shown in Fig. 3-16.

3-16 Worked example

Cartesian coordinate axes are drawn in a body. They are also the principal axes for the following stresses which are applied simultaneously to the body.

(i) A pure shear of principal values ± 5 dyn/cm² in the XZ plane, the positive stress being in the x direction.

(ii) A pure shear of principal values ± 3 dyn/cm² in the XY plane, the positive stress being in the y direction.

(iii) A hydrostatic pressure of 2 dyn/cm².

Determine

(a) the values of the principal stresses,
(b) the components of the principal stresses representing the deviatoric stress,
(c) the components of the principal stresses representing the dilatational stress, and
(d) the magnitude and direction of the force acting across a plane of unit area whose outward normal lies in the YZ plane between the positive y and z directions and making an angle of $30°$ with the y axis.

(a) The three stresses corresponding to (i), (ii), and (iii) above are shown in Fig. 3-17(a). The resultant principal stresses are given by

$$\sigma_x = 5 - 3 - 2 = 0$$

$$\sigma_y = 3 - 2 = 1 \ \text{dyn/cm}^2$$

$$\sigma_z = -5 - 2 = -7 \ \text{dyn/cm}^2$$

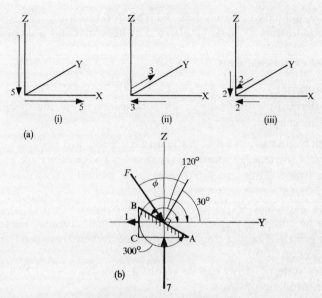

Fig. 3-17 Worked example 3-16

(b) The deviatoric stress will arise from the pure shears. Hence, from Fig. 3-17(a),

$$\sigma_{xd} = 5 - 3 = 2 \text{ dyn/cm}^2$$
$$\sigma_{yd} = 3 \text{ dyn/cm}^2$$
$$\sigma_{zd} = -5 \text{ dyn/cm}^2$$
$$\sigma_{xd} + \sigma_{yd} + \sigma_{zd} = 0$$

confirming that this is a deviatoric stress.

(c) The dilatational stress will arise from the hydrostatic pressure; thus $\sigma_d = -2 \text{ dyn/cm}^2$.

(d) Since σ_x for the resultant stress is zero, all stress components lie in the YZ plane and Eqns. (3-8) and (3-9) can be used. From Eqn. (3-9), if ϕ is the angle between the normal to the plane and the direction of action of the force [Fig. 3-17(b)],

$$\tan (30° + \phi) = -7 \tan 30°$$

giving $\qquad 30° + \phi = 103°54' \quad \text{or} \quad 283°54'$

From Fig. 3-17(b), since F must act on the unshaded side of the plane, $30 + \phi$ must lie between 0 and 120° or between 300 and 360°. Therefore $\phi = 73°54'$. From Eqn. (3-8)

$$\frac{F^2}{\sigma_z^2} \tan^2 (30° + \phi) + \frac{F^2}{\sigma_x^2} = \tan^2 (30° + \phi) + 1$$

giving $F = \pm 3 \cdot 6$ dyn. From Fig. 3-17(b), for the triangular wedge ABC to be in equilibrium, F must be directed as indicated and hence is negative.

Relevant exercises: Nos. 3-6 to 3-8.

3-17 Summary

We have derived the following major conclusions about the specification of stress in this chapter.

(i) With respect to any set of cartesian axes drawn in a body, six quantities, three normal (or tensile) and three shear stress components are necessary to define the stress. Thus stress, like strain, is a second rank symmetric tensor (Section 3-5).

(ii) One particular set of axes can be found with respect to which the stress can be completely specified by three normal stresses. These axes are called the principal axes of stress and the three stresses are called the principal stresses (Section 3-8).

(iii) Any state of stress can be resolved into dilatational and deviatoric components (Section 3-15).

  (a) The component of each principal stress due to the dilatational component is given by Eqn. (3-12).
  (b) The components of each principal stress due to the deviatoric component are given by Eqns. (3-13).
  (c) The deviatoric stress can be broken down into various combinations of pure shears, some of which are illustrated in Fig. 3-16.

(iv) Two principal stresses of equal magnitude but opposite sign, with the third principal stress zero, represent a pure shear. On axes lying at 45° to the non-zero principal axes and in their plane, the normal stresses are zero, and the shear stress is equal to the principal stress (Section 3-12).

EXERCISES

Sections 3-1 to 3-6

3-1 A plane of area A forms part of a surface drawn within a block of material. If F is the outward acting contact force exerted on this plane by the material lying outside the surface and θ is the angle between this force and the outward normal to the plane, complete the following table.

A (cm²)	F (dyn)	θ (degrees)	Normal stress (dyn/cm²)	Shear stress
10	30	10		
	25	40	0·958	
5	8·90		−1·52	

3-2 The contact forces between the faces of a rectangular parallelepiped and its surrounding material are as shown in Fig. 3-18, equal and opposite forces acting on opposite faces. Given that $\sigma_{yy} = 0 \cdot 433$ determine σ_{xx}, σ_{zz}, σ_{xy}, σ_{xz}, σ_{yz}, F, x, α, β, γ, θ, and ϕ. (Forces in dyn, length in cm, stress in dyn/cm².)

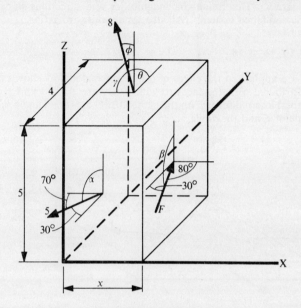

Fig. 3-18 Exercise 3-2

Sections 3-7 to 3-9

3-3 A force F is uniformly distributed over the plane end of a cylindrical rod of radius r. (This plane end is normal to the axis of the rod.) Calculate the normal and shear stresses on a plane whose normal makes an angle θ with the rod axis. Determine the values of θ at which these stresses have their maximum value and calculate these maxima.

3-4 Cartesian coordinate axes OX, OY, OZ define respectively the edges OA ($= 17 \cdot 32$ cm), OB ($= 20$ cm), OC ($= 10$ cm) of a rectangular block of material. This block is made in two parts welded together along a plane which contains the diagonal AC and whose normal lies in the xz plane. The welded joint will break if the shear stress applied to it exceeds 3×10^8 dyn/cm² or the normal stress exceeds 6×10^8 dyn/cm². A force is distributed uniformly over the face passing through A and normal to the x axis. The line of action of this force lies in the xz plane and makes an angle of $77°54'$ with the x axis. (This angle is formed by the rotation of a line from the positive x axis towards the negative z axis.) The force acts towards the negative x direction and its total magnitude is $1 \cdot 51 \times 10^{11}$ dyn. An equal and opposite force is distributed over the face passing through O and normal to the x axis. Find the magnitude and direction of forces distributed uniformly over the faces normal to the z direction, the magnitude and direction of these forces being such as to restore equilibrium without fracturing the joint.

3-5 Stresses of 3×10^8 dyn/cm^2 and 5×10^8 dyn/cm^2 are applied along the x and y directions respectively in a sheet of material. A further pair of perpendicular stresses of 4×10^8 dyn/cm^2 and 6×10^8 dyn/cm^2 are applied along axes at 30° to the original pair. (The 4×10^8 stress is rotated through 30° anticlockwise from the 3×10^8 stress.) Determine the magnitudes and directions of the principal stresses of the combined system. (All stresses are outward acting.)

Sections 3-10 to 3-16

3-6 Forces are applied to the edges of a plate of material as shown in Fig. 3-19. (They are distributed over the faces to which they are applied and are normal to them.) The total internal force F on the plane DBEG acts at an angle α. Determine the magnitude of F and the angle α.

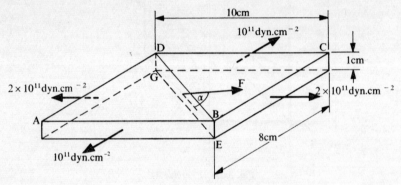

Fig. 3-19 Exercise 3-6

3-7 Principal stresses $\sigma_x = 2$ dyn/cm^2, $\sigma_y = 0$ dyn/cm^2, $\sigma_z = -1$ dyn/cm^2, are applied to a sheet of material. (The y axis is normal to the face of the sheet.) The normal to a unit plane drawn in the material lies in the xz plane and makes an angle θ with the positive x axis (by an anticlockwise rotation of that axis). The force F acting on this plane makes an angle ϕ with the positive x axis (again by anticlockwise rotation of that axis). Determine the values of ϕ and F and state whether F is inward or outward acting for values of θ of 20°, 54°44′, and 70°.

3-8 Principal stresses $\sigma_x = 4$ dyn/cm^2, $\sigma_y = 6$ dyn/cm^2, $\sigma_z = -1$ dyn/cm^2, are applied to a block of material. Determine
   (a) their components due to the dilatational component of the stress,
   (b) the components due to the deviatoric component.
   Determine three different combinations of pure shears which are equal to this deviatoric component, and draw diagrams to illustrate them.

4

Relationship between states of stress and strain

4-1 Introduction

We are now able to specify the state of strain in a body which has suffered a known deformation. We can also specify the state of stress if we know the forces acting on the body. In making these specifications, we do not need to know anything about the material from which the body is made. For example, if a load is hung on the end of a long vertical wire, the state of stress depends only on the magnitude of the load and the diameter of the wire. If these are the same for different wires, then the states of stress will be the same, whatever the materials from which the wires are made. Similarly, we can calculate the states of strain knowing only the initial and final diameters and lengths of the wires. However, if a number of wires of the same length and diameter, but made from different materials, are strained by hanging the same load on each, their changes in length and diameter will be different. That is, the properties of the material determine the magnitude of the state of strain produced by a given state of stress.

In this chapter, we consider how to calculate the state of strain when the state of stress is given (and vice versa), and determine what we need to know about the properties of the material in order to make this calculation.

4-2 Isotropically deformable bodies

For simplicity, we consider only one particular class of deformable bodies, and apply the restriction to small strain introduced in Chapter 2. These restrictions would not be necessary in a completely general treatment, but the mathematical difficulties would then be beyond the scope of this book.

The bodies we consider are called *isotropically deformable*. The meaning of this term can be made clear by a simple experiment. Suppose a series of strips, all of the same cross-sectional area and length, are cut from a body, each strip being cut so that its axis corresponds to a different direction in the body. Let the strips be hung vertically with the same load attached to each (i.e., the state of stress is the same for all). If the changes in cross-sectional area and length (the state of strain) are the same for each strip, that is, if the state of strain is independent of the initial orientation of the strip in the body, then the body is said to be isotropically deformable. In other words, the response of such a body to a deforming stress is the same in all directions in the body.

4-3   Dilatational stress on an isotropically deformable body

We defined dilatational stress in Section 3-11 as one for which the three principal stresses were equal. We saw that, for such a stress, the magnitude of the internal force acting on a unit plane drawn in the body was independent of the direction of the plane and was always normal to it. Consider now any pair of parallel planes drawn in the body, as shown in Fig. 4-1. A and B

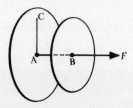

Fig. 4-1   Force on any pair of parallel planes in a body due to dilatational stress

represent two particles, one lying in each plane, such that the line joining them is normal to the planes. AC represents any line of particles lying in one of the planes. If a dilatational stress acts on the body, the force tending to displace these planes relative to each other acts along the line AB. It therefore tends to separate the planes in this direction. It might also tend to change the angle BAC and the length of AC. However, before deformation, angle BAC is a right angle whatever the direction of AC, and the force F acts along AB. Thus, since the material is isotropic, if angle BAC changes it must change by the same amount, whatever the direction of AC. This can only be so if it remains a right angle. AB will therefore change in length (and AC might), but the angle between them will remain a right angle. The lines AB and AC are therefore principal axes of strain.

Since the stress is dilatational, the force F is normal to the planes in Fig. 4-1, whatever their direction, and, since the material is isotropic, they will respond in the same way to the force F. Hence it follows that any set of mutually perpendicular lines can be chosen as principal axes, and so the strain must be dilatational (Section 2-14). We can say, therefore, that *a dilatational stress acting on an isotropically deformable body produces a dilatational state of strain.*

A dilatational stress can be represented by only one quantity—one of the principal stresses. Similarly, a dilatational strain can be represented by only one quantity—either the principal strain in dilatation or the dilatational strain (Section 2-14). Thus, the state of strain resulting from a dilatational stress can be derived from the relationship between these quantities, and so it follows that only one relationship which depends upon the nature of the material (*material-dependent relationship*) is necessary to define the response of an isotropic body to a dilatational stress. Note that, had the state of strain not been dilatational (which would have been the case if the body had been anisotropic), several such relationships might have been necessary to determine the response to a dilatational stress. This is because it would then have been necessary to determine the magnitudes and directions of the principal axes of the strain ellipsoid to specify the state of strain. Each of these would be related to the stress by a material-dependent relationship, and some of the relationships might be independent of the others.

4-4 Pure shear stress on an isotropically deformable body

Before we discuss the effect of pure shear stress, we need to make the following assumption about the way in which stress and strain are related. *Suppose a state of stress A_1 causes a state of strain B_1, and a state of stress A_2 causes a state of strain B_2. We assume that a state of stress $A_1 + A_2$ will cause a state of strain $B_1 + B_2$.* We have already seen that, provided strain is small, both stress and strain are superposable quantities, so we are now assuming that if the stress A_2 is applied alone it will produce a strain B_2, whereas if it is superimposed on another stress it will produce an increase B_2 in the state of strain. In other words, the change in the state of strain produced in a body by a given change in its state of stress is independent of the extent to which the body has already been strained (provided strain is small). We are assuming, in fact, that the elements of the stress and strain tensors are linearly related, and this is called the assumption of linearity.

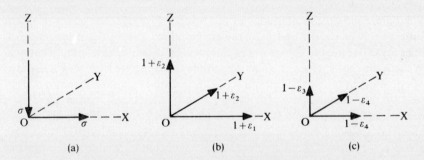

Fig. 4-2 (a) Pure shear stress, (b) Strains resulting from stress along OX, (c) Strains resulting from stress along OZ

Now, from Section 3-8, a pure shear stress can be represented by two principal stresses of equal magnitude but of opposite sign, the third principal stress being zero. The strain produced by each of these stresses acting individually can be deduced, and if the above assumption is valid they can be superimposed to give the final state of strain.

In Fig. 4-2(a), the pure shear stress is represented by principal stresses of $+\sigma$ and $-\sigma$ along OX and OZ, respectively. Let us consider the effects of these stresses on the lines of particles lying along OX, OY, and OZ, dealing first with each one separately and then considering their combined effect.

The stress $+\sigma$ along OX causes the line of particles lying along OX to change in length but, by the arguments used in the previous section, because of the isotropy of the material, this line must remain perpendicular to those particles in the YZ plane. For the same reason, although lines of particles in this plane might change in length, they must all change equally. Thus, if the lengths OX, OY, and OZ are each unity before deformation, their lengths afterwards can be written as $1 + \varepsilon_1$, $1 + \varepsilon_2$, $1 + \varepsilon_2$, respectively. This is the most general statement that we can make about the changes in length caused by the stress along OX, since it assumes nothing about the relative magnitudes, or signs, of ε_1 and ε_2. Since OX remains perpendicular to the YZ plane during deformation, it must be a principal axis of strain, as well as of stress, and the other principal axes of strain must lie in the YZ plane. The deformation at this stage is shown in Fig. 4-2(b).

For the effect of the stress $-\sigma$ along OZ, using a similar argument to the one above, we can say that the line of particles along OZ will remain perpendicular to the plane XY during deformation, that is, OZ is a principal axis of strain as well as of stress, and that the changes in length of the three lines of particles will be as shown in Fig. 4-2(c).

Now if both OX and OZ are principal axes of strain, it follows that OY must also be one. Thus, if a pure shear stress is applied to an isotropically deformable body, the principal axes of strain coincide with the principal axes of stress.

To determine the magnitudes of the principal strains, the states of strain shown in Figs. 4-2(b) and 4-2(c) must be added (using the assumption of linearity). However, before doing this we can make further simplifications. We assume that stresses of equal magnitude but opposite sign produce strains of equal magnitude but opposite sign. (Since we have already assumed that stress–strain relationships are linear, this is inevitable, unless there is a discontinuity in such relationships at the origin.) Then, if a stress of $+\sigma$ along OX causes a strain ε_1 in this direction, a stress $-\sigma$ along OZ will cause a strain $-\varepsilon_1$ along OZ. Therefore ε_1 and ε_3 are equal. Similarly, if a stress of $+\sigma$ along OX produces a strain of ε_2 along OZ, then a stress of $-\sigma$ along OZ will produce a strain of $-\varepsilon_2$ along OX. Therefore, ε_2 and ε_4 are equal.

The strains due to each stress component can now be tabulated, as below, and added to give the total strain. From Section 2-15, this represents a pure shear strain (since the strains are small).

Strain along	Due to σ along OX	Due to $-\sigma$ along OZ	Total strain
OX	ε_1	$-\varepsilon_2$	$\varepsilon_1 - \varepsilon_2$
OY	ε_2	$-\varepsilon_2$	0
OZ	ε_2	$-\varepsilon_1$	$\varepsilon_2 - \varepsilon_1$

Thus, we can say that *a pure shear stress applied to an isotropically deformable material having a linear stress–strain relationship will produce a pure shear strain on the same principal axes*. Now, if the principal axes are known, both pure shear stresses and pure shear strains can be represented by single numbers. Thus, the state of strain resulting from a pure shear stress can be determined from the relationship between these numbers, and so only one material-dependent relationship is necessary to define the response of this material to a pure shear stress.

Note that we used the assumption of isotropy several times in the above argument, and so the conclusion would not hold for an anisotropic material. For such a material, we would need to know all the parameters of the strain ellipsoid to specify the resulting strain, and the relationship between each parameter and the stress would have to be known to make the calculation.

4-5 Deviatoric stress on an isotropically deformable body

From Section 3-14, a deviatoric stress can be regarded as being composed of pure shears on two of the principal planes. We have seen above that a pure shear stress will produce a pure shear strain. Hence, the resultant state of strain will be a superposition of pure shears, and, from Section 2-17, this is a deviatoric strain. Thus, a deviatoric stress will produce a deviatoric strain on the same principal axes. Furthermore, since the material is isotropic, the shear strain resulting from each of the pure shear stresses into which the deviatoric stress is resolved can be determined from the same material-dependent relationship—that between a pure shear stress and a pure shear strain. Again the assumption of isotropy has entered into the above conclusion. In general, for an anisotropic material, the principal axes of stress and strain will not coincide, the strain will not be entirely deviatoric, and more than one material-dependent relationship will be necessary to relate the stress and strain.

4-6 Any state of stress on an isotropically deformable body

From Section 3-15, any state of stress can be resolved into deviatoric and dilatational components. From Section 4-3, the dilatational component will produce only a dilatational strain, and this can be determined knowing only a single material-dependent relationship. From Section 4-5, the deviatoric component will produce only a deviatoric strain on the same principal axes, and this can be determined knowing only a single material-dependent relationship. Providing the strains are small and the assumption of linearity is valid, these two strain components can then be superimposed to give the state of strain resulting from the applied stress. This will be on the same principal axes as the applied stress, and will have been determined knowing only two material-dependent relationships.

Hence, for any state of stress, the principal axes of stress and strain will be coincident, and the magnitude of the strain can be determined knowing only two material-dependent relationships—that between dilatational stress and dilatational strain, and that between a pure shear stress and a pure shear strain.

This is a somewhat surprising conclusion. There are innumerable ways of applying a stress to a body, and each of these ways produces a different kind of deformation. It would appear at first sight that different information about the material might be necessary to relate each stress to each strain. However, our analysis in the preceding chapters has shown that, providing the strains are small, any stress or strain may be represented by a dilatation combined with pure shears. It therefore follows from the arguments in this chapter that, provided the material is isotropically deformable and stress and strain are linearly related, then only two material-dependent relationships are necessary to determine the deformation produced by any applied stress. One of these gives information about the forces necessary to change the volume of the material, keeping its shape constant, i.e., to increase the distance apart of the molecules. This is called the *bulk modulus* (Section 4-8). The other gives information about the forces necessary to change the shape of the material, keeping its volume constant, i.e., to slide planes of molecules over each other keeping their distance apart the same, and is called the *shear modulus* (Section 4-8).

4-7   Nature of the material-dependent relationships

Our analysis of stress and strain has now reached a point where we might start the experimental study of materials. Hitherto, the value of experiments was limited because it was not clear which types of stress and strain would yield the most useful results. Also, it was not certain whether different methods of deforming a body would supply different information about the material, or just supply the same information in different form. We have now resolved these difficulties, but another difficulty remains. Stresses are the internal forces acting across planes of unit area drawn in the body; strains are determined from the lengths of the principal axes of the ellipsoids which result from drawing spheres of unit radius in the body before deformation. We cannot measure either of these quantities directly; we can only measure the forces applied to the surface of a body and the changes in its dimensions.

While it might be possible to deform a body, in uniform dilatation or pure shear, so that the stress and strain could be calculated directly from the forces applied to its boundaries and the changes in its dimensions, such experiments are not easy. It is simpler to make assumptions about the nature of the stress–strain relationships, and use these to calculate the deformations produced by stress systems which can be easily applied experimentally. If we then compare the experimental results with the calculated results, we will test the validity of the assumptions. Let us make the following assumptions (the second is the assumption of linearity introduced in Section 4-3; it is re-stated here to gather relevant information together):

1. *The strain is completely defined by the stress.*
2. *The strain is proportional to the stress.*

The first assumption implies that, if we know the stress, we have sufficient information to calculate the strain. For example, the strain produced by a given stress would be the same, whether the stress was applied and held constant for a very long time, or whether it was applied only for a fraction of a second. According to the assumption, information of this nature is irrelevant. (It must, however, be realized that the assumption is not intended to imply that the strain produced by a given stress is independent of the temperature. Throughout this book it is assumed that temperature is constant.) This assumption is known as the assumption of *perfect elasticity*, and a material obeying it is known as a *perfectly elastic material*.

The second assumption implies that, for pure shear or dilatation, a graph of stress against strain would be a straight line through the origin. This assumption is known as *Hooke's law* and materials obeying it are called *Hookeian materials*. If the relationships between stress and strain are linear, then each relationship can be defined by a single number, the elastic modulus, giving the slope of the line.

4-8 Elastic moduli

We have now seen that the strain produced in an isotropic, perfectly elastic, Hookeian material can be derived for any state of stress knowing only two material-dependent constants—the bulk and shear moduli of elasticity. They are defined as follows.

Bulk modulus of elasticity k

Consider a dilatational stress applied to a body. Let it be specified by the principal stress σ_d. The strain produced is entirely dilatational and is specified by the dilatational strain Δ, which is defined in Section 2-14 as the change in volume per unit volume. Then the bulk modulus k is given by

$$k = \sigma_d/\Delta \qquad (4\text{-}1)$$

Since the principal strain in dilatation $\varepsilon_d = \Delta/3$, then

$$\sigma_d = 3\varepsilon_d k \qquad (4\text{-}2)$$

Note that the bulk modulus of elasticity often appears in thermodynamical equations. It is then defined as

$$k = -V\,dP/dV \qquad (4\text{-}3)$$

where V is the volume and P the pressure. Since the pressure is hydrostatic, it is equal in magnitude to the principal stress, and dV/V is the dilatational strain Δ. Thus Eqns. (4-1) and (4-3) are identical except for the negative sign. This arises because inward acting pressures are taken as positive in thermodynamics, whereas outward acting stresses are positive in elasticity. Since an outward acting stress causes an increase in volume in all substances, both equations lead to positive values of bulk modulus, and moduli determined according to Eqn. (4-1) can be used in thermodynamical equations.

Shear modulus of elasticity μ

Consider a pure shear stress applied to a body. Let it be specified by the positive principal stress σ. The pure shear strain which is produced is specified by the angle of shear γ. (The particular angle of shear referred to is the change in angle between two lines which are initially perpendicular, and which do not change in length under shear.) The shear modulus μ is given by

$$\mu = \sigma/\gamma \qquad (4-4)$$

From Section 2-11, the two lines between which the angle of shear is measured are at 45° to the principal axes, and the shear strain between these lines is of the same magnitude as the principal strain. Also, from Section 2-3, the shear strain is equal to half the angle of shear. Thus, if ε is the magnitude of the principal strain, $\varepsilon = \frac{1}{2}\gamma$, and hence

$$\sigma = 2\varepsilon\mu \qquad (4-5)$$

4-9   Simple shear

So far, in considering the relationship between a shear stress and strain, we have considered only pure shear. In Section 3-13 we showed that, if a simple shearing force is applied to a body, then the system of forces which exists at equilibrium constitutes a pure shear, the additional forces required to maintain equilibrium being generated by rigid body movement. Thus, the strain must be a pure shear plus a rigid body movement. Now, from the definition of a simple shear stress, the directions in space of planes of particles at 45°

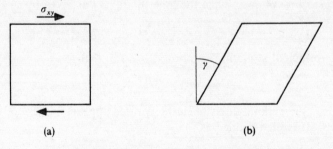

(a)   (b)

Fig. 4-3   Simple shear stress and strain

to the principal axes are fixed. Thus, the rigid body rotation must satisfy this condition and, from Section 2-15, such a deformation is a small simple shear strain.

Hence, the simple shear stress shown in Fig. 4-3(a) will produce the simple shear strain shown in Fig. 4-3(b). Furthermore, since σ_{xy} is equal in magnitude to the principal stress, and the angle of shear γ is that used in the definition of the shear modulus, σ_{xy} and γ are related by the shear modulus μ as in Eqn. (4-4).

4-10 Worked examples

1. Forces are applied to a block of material as illustrated in Fig. 4-4. The forces are given in 10^9 dyn and the dimensions in cm. Equal and opposite forces are applied to the other faces to maintain equilibrium. Under the

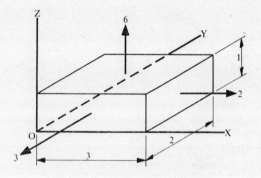

Fig. 4-4 Worked example 4-10(1)

action of these forces the edges in the x direction increase in length by 0·003 cm. Determine the changes in length of the other edges and the bulk modulus of the material.

Each of the forces acts in a direction perpendicular to the face to which it is applied; therefore there is no tangential component acting on any face. Also, the applied forces are mutually perpendicular. Hence, the directions of these forces are the directions of the principal axes of stress. The magnitudes of the principal stresses are found by dividing the magnitude of the force by the area of the face on which it acts. Thus

$$\sigma_x = 2 \times 10^9/2 = 10^9 \text{ dyn/cm}^2$$
$$\sigma_y = 3 \times 10^9/3 = 10^9 \text{ dyn/cm}^2$$
$$\sigma_z = 6 \times 10^9/6 = 10^9 \text{ dyn/cm}^2$$

Since the principal stresses are equal, the stress is dilatational, and so, by Section 4-3, the strain must also be dilatational. The three principal strains are therefore equal. Since the edges in the x direction increase in length by 0·003 cm

$$\varepsilon_x = 0{\cdot}003/3 = 10^{-3}$$

and ε_y and ε_z must be equal to this value. The dilatational strain is therefore equal to 3×10^{-3} and the bulk modulus equals $10^9/(3 \times 10^{-3}) = 3{\cdot}3 \times 10^{11}$ dyn/cm². The edge in the y direction increases in length by 2×10^{-3} cm, and that in the z direction by 1×10^{-3} cm.

2. Forces are applied to a block of elastically isotropic material as shown in Fig. 4-5. The forces are given in 10^8 dyn and the dimensions in cm. Equal

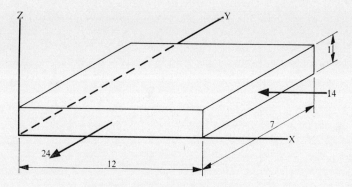

Fig. 4-5 Worked example 4-10(2)

and opposite forces are applied to the other faces to maintain equilibrium. If the shear modulus of the material is 2×10^{11} dyn/cm², determine the change in length of the sides of the sides of the block.

Using the same arguments as in the previous example, the principal axes of stress are in the directions of the coordinate axes. The principal stresses are

$$\sigma_x = -14 \times 10^8/7 = -2 \times 10^8 \text{ dyn/cm}^2$$
$$\sigma_y = 24 \times 10^8/12 = 2 \times 10^8 \text{ dyn/cm}^2$$
$$\sigma_z = 0$$

This state of stress is a pure shear and so, by Section 4-4, the strain must also be a pure shear on the same principal axes. Thus, the angle of shear is found by dividing the positive principal stress by the shear modulus, and is equal to

$$(2 \times 10^8)/(2 \times 10^{11}) = 10^{-3}$$

The principal strains are equal to half the angle of shear and are $\pm 0{\cdot}5 \times 10^{-3}$. Thus the x direction decreases in length by $0{\cdot}5 \times 10^{-3} \times 12 = 6 \times 10^{-3}$ cm and the y direction decreases in length by $0.5 \times 10^{-3} \times 7 = 3.5 \times 10^{-3}$ cm. The z direction is unchanged in length.

3. A block of material and the forces acting on it are shown in Fig. 4-6. The forces are given in 10^9 dyn and the dimensions in cm. If the edges in the x direction increase in length by 0·018 cm under the action of these forces, determine the shear modulus and the changes in length of the other edges.

Using the same arguments as in the previous examples, the principal

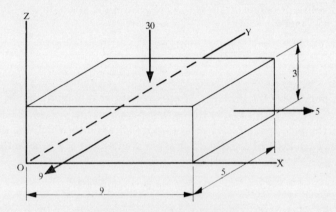

Fig. 4-6 Worked example 4-10(3)

axes of stress are in the directions of the coordinate axes. The magnitudes of these stresses are given by dividing the force by the area on which it acts. Thus

$$\sigma_x = 10^9/3 \ \text{dyn/cm}^2$$
$$\sigma_y = 10^9/3 \ \text{dyn/cm}^2$$
$$\sigma_z = -2 \times 10^9/3 \ \text{dyn/cm}^2$$

The sum of these stresses is zero and so the stress is deviatoric. It can therefore be represented by two pure shears. Furthermore, if these are chosen so that only one of them changes the length in the x direction, the shear modulus can be found from the ratio of this shear to the change in length in the x direction, which is given. This division into two shears can be done by inspection and is shown in Fig. 4-7.

If the shear modulus is μ, the angle of shear due to the first of these pure shears, which is the one causing a change in length in the x direction, is given by the positive principal stress divided by μ. Thus,

$$\text{Angle of shear} = 10^9/3\mu$$

and so the principal strains are $\pm 10^9/6\mu$.

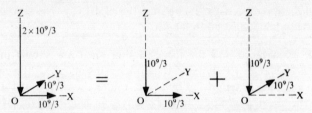

Fig. 4-7   Analysis of deviatoric stress: worked example 4-10(3)

The positive principal strain is ε_x which, from the given dimension change, is $0{\cdot}018/9 = 0{\cdot}002$.

Thus $10^9/6\mu = 0{\cdot}002$

and $\mu = 8{\cdot}3 \times 10^{10} \text{ dyn/cm}^2$

Since in pure shear the principal strains are of equal magnitude but opposite sign, the component of ε_z due to the first pure shear is $-0{\cdot}002$. The magnitudes of the principal stresses of both pure shears into which the deviatoric stress has been divided are equal; thus the component of ε_z due to the second shear is also $-0{\cdot}002$. Therefore

$$\varepsilon_z = -0{\cdot}004 \quad \text{and} \quad \varepsilon_y = 0{\cdot}002$$

Hence, the edges parallel to the y direction increase in length by $5 \times 0{\cdot}002$, or $0{\cdot}01$ cm, and the edges parallel to the z direction decrease in length by $0{\cdot}012$ cm.

4. A block of material and the forces acting on it are shown in Fig. 4-8. The forces are given in 10^9 dyn and the dimensions in cm. Given that edges

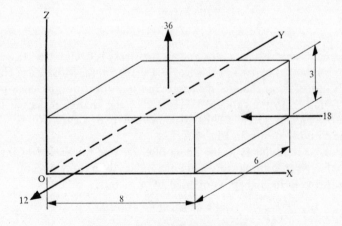

Fig. 4-8   Worked example 4-10(4)

in the x direction contract by 0.014 cm and those in the y direction extend by 0.0043 cm, calculate the change in length of edges parallel to the z direction, and the shear and bulk moduli.

As in the previous examples, the principal axes of stress are in the directions of the coordinate axes and are given by

$$\sigma_x = -10^9 \text{ dyn/cm}^2$$
$$\sigma_y = 10^9/2 \text{ dyn/cm}^2$$
$$\sigma_z = 3 \times 10^9/4 \text{ dyn/cm}^2$$

The principal stresses of the dilatational component are given by

$$\sigma_d = (-1 + 0.5 + 0.75) \times 10^9/3 = 10^9/12 \text{ dyn/cm}^2$$

Thus, if k is the bulk modulus, the principal strains of the dilatational component, from Eqn. (4-2), are given by

$$\varepsilon_d = 10^9/36k$$

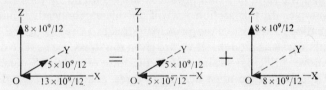

Fig. 4-9 Analysis of deviatoric stress: worked example 4-10(4)

The principal values of the deviatoric component of the stress are found by subtracting σ_d from σ_x, σ_y, and σ_z, in turn, giving

$$\sigma_{xd} = -13 \times 10^9/12 \text{ dyn/cm}^2$$
$$\sigma_{yd} = 5 \times 10^9/12 \text{ dyn/cm}^2$$
$$\sigma_{zd} = 8 \times 10^9/12 \text{ dyn/cm}^2$$

One of the ways in which this can be resolved into two pure shears is shown in Fig. 4-9.

If μ is the shear modulus, the angle of shear due to the first of these pure shears is $5 \times 10^9/12\mu$, and that due to the second is $8 \times 10^9/12\mu$. Thus, the principal strains due to the first are $\pm 5 \times 10^9/24\mu$, and those due to the second are $\pm 8 \times 10^9/24\mu$. Superimposing these gives the principal strains of the deviatoric component as

$$\varepsilon_{xd} = -13 \times 10^9/24\mu$$
$$\varepsilon_{yd} = 5 \times 10^9/24\mu$$
$$\varepsilon_{zd} = 8 \times 10^9/24\mu$$

Hence the principal values of the total strain are

$$\varepsilon_x = \varepsilon_d + \varepsilon_{xd} = 10^9/36k - 13 \times 10^9/24\mu$$

$$\varepsilon_y = \varepsilon_d + \varepsilon_{yd} = 10^9/36k + 5 \times 10^9/24\mu$$

$$\varepsilon_z = \varepsilon_d + \varepsilon_{zd} = 10^9/36k + 8 \times 10^9/24\mu$$

From the information given, $\varepsilon_x = -1\cdot4 \times 10^{-2}/8$ and $\varepsilon_y = 4\cdot3 \times 10^{-3}/6$. Substituting these in the first two of the above equations and solving for k and μ gives $\mu = 3\cdot03 \times 10^{11}$ and $k = 8\cdot84 \times 10^{11}$ dyn/cm². Substituting these values of k and μ in the third equation gives $\varepsilon_z = 0\cdot00113$, and thus the edges parallel to the z direction extend by $0\cdot00339$ cm.

Relevant exercises: Nos. 4-1 and 4-2.

4-11   Relationship between principal stresses and strains

The system of stresses applied in Example 4 above had both dilatational and deviatoric components. Since any state of stress can be resolved into these components, the state of strain produced can be found by the method used in this example. In this section we will do this algebraically, and derive a set of equations expressing the principal stresses in terms of the principal strains and the elastic moduli. If we are given the stress, we have to determine three unknown principal strains to find the state of strain. Similarly, if we are given the strain, we must determine three unknown principal stresses. Hence, we require a set of three simultaneous equations.

The state of stress is defined by the principal stresses σ_x, σ_y, and σ_z, and

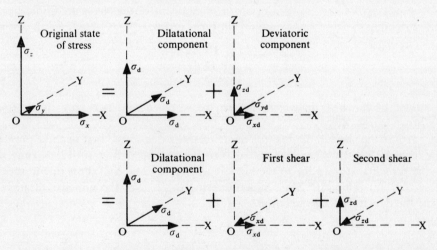

Fig. 4-10   Analysis of stress into a dilatation and two shears to determine the relationship between principal strains and principal stresses

we can analyse this state into a dilatational component and two pure shears as follows (Fig. 4-10).

The principal stresses of the dilatational component are given by

$$\sigma_d = \tfrac{1}{3}(\sigma_x + \sigma_y + \sigma_z) \tag{4-6}$$

The principal stresses of the deviatoric component are given by

$$\sigma_{xd} = \sigma_x - \sigma_d \tag{4-7a}$$

$$\sigma_{yd} = \sigma_y - \sigma_d \tag{4-7b}$$

$$\sigma_{zd} = \sigma_z - \sigma_d \tag{4-7c}$$

The deviatoric component can be regarded as being composed of two shears:
(1) in the principal plane normal to the z axis and comprising σ_{xd} along the x axis and $-\sigma_{xd}$ along the y axis, and
(2) in the principal plane normal to the x axis and comprising σ_{zd} along the z axis and $-\sigma_{zd}$ along the y axis.

Thus

$$\sigma_{yd} = -(\sigma_{xd} + \sigma_{zd}) \tag{4-8}$$

which would be expected for a deviatoric stress.

We can find the principal strains of the dilatational strain component from the dilatational component of the stress. From Eqn. (4-2), they are given by

$$\varepsilon_d = \sigma_d/3k \tag{4-9}$$

We can find the principal strains of the deviatoric strain component from the pure shear stresses. From the first shear stress and Eqn. (4-5)

$$\varepsilon_{xd} = \sigma_{xd}/2\mu \tag{4-10a}$$

From the second shear stress

$$\varepsilon_{zd} = \sigma_{zd}/2\mu \tag{4-10b}$$

Since these principal values, together with ε_{yd}, must form a deviatoric strain, then

$$\varepsilon_{yd} = -(\varepsilon_{xd} + \varepsilon_{zd}) \tag{4-10c}$$

The principal stresses and strains can now be related by substituting from Eqns. (4-9) and (4-10) in Eqn. (4-7)

$$\sigma_x = 3k\varepsilon_d + 2\mu\varepsilon_{xd} \tag{4-11a}$$

$$\sigma_y = 3k\varepsilon_d + 2\mu\varepsilon_{yd} \tag{4-11b}$$

$$\sigma_z = 3k\varepsilon_d + 2\mu\varepsilon_{zd} \tag{4-11c}$$

We now have to express ε_d and ε_{xd}, etc., in terms of the total principal strains ε_x, ε_y, and ε_z.

Now
$$\varepsilon_d = \tfrac{1}{3}(\varepsilon_x + \varepsilon_y + \varepsilon_z) \tag{4-12}$$

and
$$\varepsilon_{xd} = \varepsilon_x - \varepsilon_d \tag{4-13a}$$

$$\varepsilon_{yd} = \varepsilon_y - \varepsilon_d \tag{4-13b}$$

$$\varepsilon_{zd} = \varepsilon_z - \varepsilon_d \tag{4-13c}$$

Substituting for ε_{xd}, etc., in Eqn. (4-11) from (4-13) gives

$$\sigma_x = (3k - 2\mu)\varepsilon_d + 2\mu\varepsilon_x \tag{4-14}$$

with similar equations for σ_y and σ_z.

Substituting for ε_d in Eqn. (4-14) from (4-12) and writing

$$\lambda = k - \tfrac{2}{3}\mu \tag{4-15}$$

gives
$$\sigma_x = (\lambda + 2\mu)\varepsilon_x + \qquad \lambda\varepsilon_y + \qquad \lambda\varepsilon_z \tag{4-16a}$$

$$\sigma_y = \qquad \lambda\varepsilon_x + (\lambda + 2\mu)\varepsilon_y + \qquad \lambda\varepsilon_z \tag{4-16b}$$

$$\sigma_z = \qquad \lambda\varepsilon_x + \qquad \lambda\varepsilon_y + (\lambda + 2\mu)\varepsilon_z \tag{4-16c}$$

We make the substitution of Eqn. (4-15) merely for convenience in writing down the equations. Since λ is expressed in terms of k and μ, it is another means of stating one of the moduli needed to define the elastic properties of a material. However, it does not have the clear physical interpretation of k and μ. λ and μ are referred to as the *Lamé constants* of the material.

The Lamé constants can thus be defined as two moduli necessary to relate the principal stresses and strains in an isotropically deformable material. The first, given by the symbol λ, is equal to the difference between the bulk modulus and two-thirds of the shear modulus; the second is equal to the shear modulus.

From Eqns. (4-16), we can determine the principal stresses if the principal strains are given. In order to determine the principal strains when the principal stresses are given, we can derive equations by solving the three simultaneous equations (4-16) for ε_x, ε_y, and ε_z. The solutions are

$$2\mu(3\lambda + 2\mu)\varepsilon_x = 2(\lambda + \mu)\sigma_x \qquad - \lambda\sigma_y \qquad - \lambda\sigma_z \tag{4-17a}$$

$$2\mu(3\lambda + 2\mu)\varepsilon_y = \qquad - \lambda\sigma_x + 2(\lambda + \mu)\sigma_y \qquad - \lambda\sigma_z \tag{4-17b}$$

$$2\mu(3\lambda + 2\mu)\varepsilon_z = \qquad - \lambda\sigma_x \qquad - \lambda\sigma_y + 2(\lambda + \mu)\sigma_z \tag{4-17c}$$

The coefficients of the terms in both sets of Eqns. (4-16) and (4-17) are arranged according to the same pattern, which should be carefully noted. Coefficients on the left-hand side are all the same; those on the right-hand side form an array. The coefficients lying on the principal diagonal of this array are identical; the remaining six are also identical. Hence, each set of equations comprises only three different terms, the left-hand side, the diagonal, and the off-diagonal. Once these are known, the complete set can be written down, or, given one equation, the remaining two can be deduced.

We will explore the properties of these equations in worked examples, but the form of Eqns. (4-17) reveals one feature of paramount importance in studies of stress and strain. Each one of the principal strains, ε_x, ε_y, ε_z, is influenced by all three principal stresses, not just that along its own axis. Thus, if a stress is applied to a body along one direction only, the body will be in a three-dimensional state of strain. Similarly, if a one-dimensional strain is required, then stresses must be applied along all three dimensions. It is this characteristic which makes the response of an isotropic body to stress different from its response to other physical stimuli. For example, if an electric potential gradient is applied to such a body, then the current will flow in the direction of this gradient. Thus, considerable progress can be made in the study of electrical phenomena by one-dimensional treatments—a simplification not possible with elasticity. It is because of this difference in experimental response that stress is a second rank tensor, whereas quantities such as potential gradient are vectors.

4-12 Relationship between non-principal stresses and strains

If we know the principal strains, then using Eqns. (4-16) we can calculate the principal stresses. Similarly, if we know the principal stresses, we can calculate the principal strains using Eqns. (4-17). If the state of stress is given with reference to coordinate axes which do not lie along the principal directions, it is always possible, in principle, to determine these directions and hence the principal stresses. We can then use the above equations to find the principal strains, and the strain components with respect to the original coordinate axes can then be calculated if necessary. But there is a more direct method, given below. We consider the case in which the stress components are given and the strain components are required. The method could equally well be used to find the stress to produce a given state of strain.

If the coordinate axes do not lie along the principal directions, we need six quantities to specify the stress—σ_{xx}, σ_{yy}, σ_{zz}, σ_{xy}, σ_{zx}, and σ_{yz}. This stress can be applied in two stages, as shown in Figs. 4-11(a)–(c). (For clarity, the stress components acting on the hidden faces in this diagram have been omitted.) First, the normal stresses, σ_{xx}, etc., can be applied [Fig. 4-11(b)], and then the shear stresses [Fig. 4-11(c)]. If we can find the states of strain for each of these states of stress then, for small strains, they can be superimposed to give the final state of strain.

We can easily determine the state of strain caused by the normal stresses acting on their own from Eqns. (4-17), since the principal directions of such a state of stress would lie along the coordinate axes. The principal strains so obtained lie along the coordinate axes, and so are extensional strains, ε_{xx}, etc., in these directions.

Consider now the stress σ_{xy} acting on its own [Fig. 4-11(d)]. From Section 3-8, this is a pure shear with principal stresses of magnitude $\pm\sigma_{xy}$ in directions

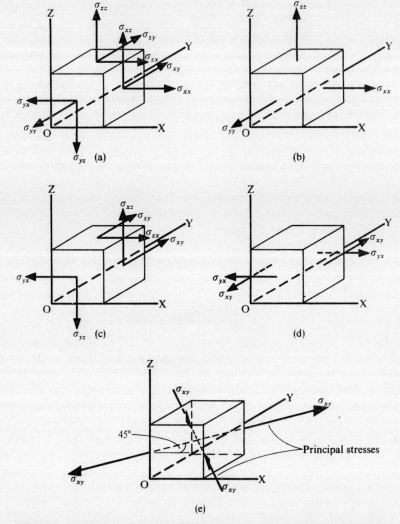

Fig. 4-11   Derivation of relationship between non-principal stresses and strains
(a) Original state of stress, (b) 1st stage to be applied, (c) 2nd stage to be applied, (d) Shear stress σ_{xy} applied alone, (e) Shear stress σ_{xy} referred to principal axes

at 45° to the x and y coordinate axes [Fig. 4-11(e)]. The strain produced by this stress must therefore be a pure shear on the same principal axes, and so σ_{xy} will produce only strain components in the XY plane (i.e., only ε_{xx}, ε_{yy}, and ε_{xy} can be non-zero). Now, from Section 2-11, a pure shear strain does not alter the lengths of lines at 45° to the principal axes. Since, in this case, these are lines parallel to the directions of the x and y coordinate axes, ε_{xx} and ε_{yy} must both be zero, and only ε_{xy} can be non-zero. We can find the angle of shear between the x and y axes from the shear modulus and the principal stress, using Eqn. (4-4), and determine the shear strain ε_{xy}, which is equal to half the magnitude of this angle. In a similar way, we can find ε_{xz} and ε_{yz} from σ_{xz} and σ_{yz}, respectively, and for each of these stresses the other strain components are all zero.

Thus, we can find strain components with respect to any set of axes, knowing the stress components for the same axes, by determining the extensional strains from the normal stresses using Eqns. (4-17), and the shear strains from the shear stresses using Eqn. (4-4).

4-13   Worked examples

1. A principal stress σ_x is applied to a material with Lamé constants λ and μ, the other principal stresses being zero.
  (a) Determine the principal strains.
  (b) Calculate the principal stresses, $\sigma_y = \sigma_z = \sigma$, which must be applied to restore ε_y and ε_z to zero, and find the change they produce in ε_x.
  (c) Determine the principal stresses which must be applied if ε_y and ε_z are zero whilst ε_x is the same as in (a) above.

(a) Initially the principal stresses are σ_x, $\sigma_y = \sigma_z = 0$. To find the principal strains, substitute these values in Eqns. (4-17):

$$\varepsilon_x = \frac{(\lambda + \mu)}{\mu(3\lambda + 2\mu)}\,\sigma_x$$

$$\varepsilon_y = \varepsilon_z = \frac{-\lambda}{2\mu(3\lambda + 2\mu)}\,\sigma_x$$

(b) Principal stresses $\sigma_y = \sigma_z = \sigma$ are now to be applied, σ_x remaining unaltered, so that $\varepsilon_y = \varepsilon_z = 0$. Let the new value of ε_x be ε'_x. Since, in this case, two of the principal strains are zero, it is simpler to substitute the known stress and strain components in Eqns. (4-16):

$$\sigma_x = (\lambda + 2\mu)\varepsilon'_x$$

$$\sigma = \lambda\varepsilon'_x$$

Eliminating ε'_x between these equations gives

$$\sigma = \frac{\lambda}{\lambda + 2\mu}\,\sigma_x$$

which is the principal stress that must be applied in the y and z directions to restore the principal strains in this direction to zero. Substituting the value obtained for σ gives

$$\varepsilon'_x = \frac{\sigma_x}{\lambda + 2\mu}$$

Therefore the change in ε_x is given by

$$\varepsilon_x' - \varepsilon_x = \left[\frac{1}{\lambda + 2\mu} - \frac{(\lambda + \mu)}{(3\lambda + 2\mu)}\right]\sigma_x$$

which, on simplification, becomes

$$\varepsilon_x' - \varepsilon_x = \frac{-\lambda^2}{\mu(3\lambda + 2\mu)(\lambda + 2\mu)}\sigma_x$$

(c) Let the principal stresses which must be applied so that ε_y and ε_z are both zero, whilst ε_x is the same as in (a) above, be given by σ_x' in the x direction, and σ' in the y and z directions. These stresses must produce principal strains $\varepsilon_y = \varepsilon_z = 0$, and $\varepsilon_x = \sigma_x(\lambda + \mu)/\mu(3\lambda + 2\mu)$. Substituting these values in Eqns. (4-16) gives

$$\sigma_x' = \frac{(\lambda + 2\mu)(\lambda + \mu)}{\mu(3\lambda + 2\mu)}\sigma_x$$

$$\sigma' = \frac{\lambda(\lambda + \mu)}{\mu(3\lambda + 2\mu)}\sigma_x$$

2. The stress acting on a body is defined by the following components:

$$\sigma_{xx} = 3 \times 10^9 \text{ dyn/cm}^2 \qquad \sigma_{yy} = 5 \times 10^9 \text{ dyn/cm}^2$$

$$\sigma_{zz} = -1 \times 10^9 \text{ dyn/cm}^2 \qquad \sigma_{xy} = 2\cdot5 \times 10^9 \text{ dyn/cm}^2$$

$$\sigma_{xz} = -1\cdot5 \times 10^9 \text{ dyn/cm}^2 \qquad \sigma_{yz} = 4 \times 10^9 \text{ dyn/cm}^2$$

If the Lamé constants of the material are $\lambda = 10^{12}$ and $\mu = 3 \times 10^{11}$ dyn/cm^2, determine the strain components. If the material is in the form of a rectangular parallelepiped, with edges parallel to the coordinate axes and of length OX = 5 cm, OY = 10 cm, OZ = 3 cm, determine the change in the lengths of these sides and the change in the angle between them. Determine also the change in length of the diagonal of the parallelepiped from the origin.

From Section 4-12, ε_{xx}, etc., are given by substituting σ_{xx}, etc., in Eqns. (4-17):

$$\varepsilon_{xx} = \frac{2 \times 1\cdot3 \times 10^{12} \times 3 \times 10^9 - 10^{12} \times 4 \times 10^9}{0\cdot6 \times 3\cdot6 \times 10^{24}}$$

$$= 1\cdot76 \times 10^{-3}$$

Similarly,
$$\varepsilon_{yy} = 5\cdot09 \times 10^{-3}$$

$$\varepsilon_{zz} = -4\cdot90 \times 10^{-3}$$

ε_{xy}, etc., can be found by substituting σ_{xy}, etc., in Eqn. (4-5). This gives

$$\varepsilon_{xy} = 4{\cdot}16 \times 10^{-3} \qquad \varepsilon_{xz} = -2{\cdot}5 \times 10^{-3} \qquad \varepsilon_{yz} = 6{\cdot}67 \times 10^{-3}$$

The change in length of the edge OX is equal to $OX\varepsilon_{xx}$, and similarly for OY and OZ. Hence

OX extends by $8{\cdot}80 \times 10^{-3}$ cm

OY extends by $50{\cdot}9 \times 10^{-3}$ cm

OZ contracts by $14{\cdot}7 \times 10^{-3}$ cm

The change in angle between a pair of mutually perpendicular lines is equal to twice the shear strain between them. Shear strain is positive when a positive axis rotates towards a positive axis. Thus

the angle between OX and OY decreases by $8{\cdot}3 \times 10^{-3}$ radians

the angle between OX and OZ increases by $5{\cdot}0 \times 10^{-3}$ radians

the angle between OY and OZ decreases by $13{\cdot}3 \times 10^{-3}$ radians

To find the change in length of the diagonal we use Eqns. (2-12). One end of the diagonal lies at the origin, the other end at (x,y,z) where $x = 5$, $y = 10$, $z = 3$. If the end at the origin remains fixed during deformation, the other end is displaced to (x',y',z') in a coordinate system defined by the edges of the undeformed block, the coordinates (x',y',z') being given by Eqns. (2-12). Substituting for ε_{xx}, etc., in these equations gives

$$x' - x = 42{\cdot}9 \times 10^{-3} \text{ cm}$$
$$y' - y = 91{\cdot}7 \times 10^{-3} \text{ cm}$$
$$z' - z = 39{\cdot}5 \times 10^{-3} \text{ cm}$$

The length of the diagonal after deformation is given by

$$(x'^2 + y'^2 + z'^2)^{1/2}$$

From the values just derived, $x' - x$ is small compared with x, and similarly for y and z. Hence this length may be written

$$[(x^2 + y^2 + z^2) + 2(42{\cdot}9x + 91{\cdot}7y + 39{\cdot}5z) \times 10^{-3}]^{1/2}$$

which equals $\qquad [(x^2 + y^2 + z^2) + 2{\cdot}50]^{1/2}$

Before deformation the length of the diagonal is given by $(x^2 + y^2 + z^2)^{1/2}$ which equals $134^{1/2}$. Therefore the change in length of the diagonal on deformation is given by

$$(134 + 2{\cdot}50)^{1/2} - 134^{1/2}$$

This can be evaluated by the binomial theorem, giving the change in length as $0{\cdot}11$ cm.

3. A unit cube of material is acted upon by the forces shown in Fig. 4-12(a). F_1 and F_2 have a total resultant magnitude of 4×10^9 dyn, are directed in the XZ plane at an angle of $30°$ to the negative x direction, and are distributed over the faces to which they are applied so as to produce a state of uniform stress. An equal and opposite force is applied to the two opposite faces to maintain equilibrium.

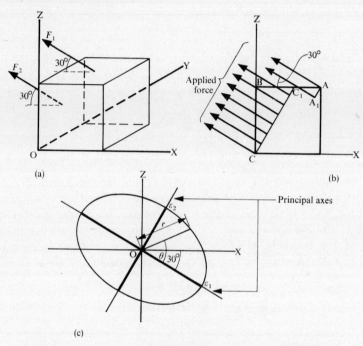

Fig. 4-12 (a) Worked example 4-15(3), (b) Stresses in XZ plane
(c) The strain ellipsoid

(a) Determine the magnitude and directions of the principal stresses.
(b) If the shear modulus is 3×10^{11} and the bulk modulus 10^{12} dyn/cm², determine the magnitudes and directions of the principal strains.
(c) Calculate the change in volume of the cube on applying the stress.
(d) Calculate the changes in length of the edges of the cube.
(e) Calculate the changes in angle between the edges of the cube.
(f) Determine the magnitudes of F_1 and F_2.

(a) The forces all lie in the XZ plane and Fig. 4-12(b) shows a section of the cube in this plane. The only force applied is a tension at $30°$ to the x axis and so this is the direction of one principal axis of stress. No forces act in a plane perpendicular to this direction, and so the other principal directions can be any two mutually perpendicular lines in this plane. For convenience, one direction will be taken as the line of intersection of this plane with the XZ plane, the other direction will then be parallel to the y axis. The applied force is distributed over the faces AB and BC. It therefore acts normally to the planes CC_1 and AA_1. Thus the principal stress is equal to the applied force divided by the area of these planes. Therefore

$$\text{principal stress} = \frac{4 \times 10^9}{\sin 30° + \cos 30°}$$

$$= 2{\cdot}93 \times 10^9 \, \text{dyn/cm}^2$$

The other two principal stresses are zero.

(b) Since $k = 10^{12}$ and $\mu = 3 \times 10^{11} \, \text{dyn/cm}^2$, from Eqn. (4-15), $\lambda = 8 \times 10^{11} \, \text{dyn/cm}^2$. Substituting λ, μ, and the principal stresses in Eqns. (4-17) gives

principal strain in direction of non-zero principal stress $= 3{\cdot}56 \times 10^{-3}$
principal strains in directions of zero principal stresses $= -1{\cdot}30 \times 10^{-3}$

(c) The dilatational strain is equal to the sum of the principal strains and is equal to the increase in volume divided by the initial volume. Therefore, the change in volume $= 0{\cdot}96 \times 10^{-3} \, \text{cm}^3$.

(d) The polar equation of the section of the strain ellipsoid lying in the XZ plane is

$$\frac{\cos^2 \theta}{(1 + 3{\cdot}56 \times 10^{-3})^2} + \frac{\sin^2 \theta}{(1 - 1{\cdot}30 \times 10^{-3})^2} = \frac{1}{r^2}$$

This is illustrated in Fig. 4-12(c) which shows the meaning of r and θ. Thus the strain in the direction of the edge lying along the x axis can be determined from

$$\frac{1}{r^2} = \frac{(\cos 30°)^2}{(1 + 3{\cdot}56 \times 10^{-3})^2} + \frac{(\sin 30°)^2}{(1 - 1{\cdot}30 \times 10^{-3})^2}$$

Neglecting squares of small quantities and solving by the binomial theorem gives $r = 1 + 2{\cdot}35 \times 10^{-3} \, \text{cm}$. Thus the extensional strain in edges parallel to the x direction is $2{\cdot}35 \times 10^{-3}$, and so these edges increase in length by $2{\cdot}35 \times 10^{-3} \, \text{cm}$. By putting θ equal to 120°, the strain in the z direction can be found, from which we find that the edges parallel to the z direction decrease in length by $0{\cdot}085 \times 10^{-3} \, \text{cm}$. Note that the values of θ used above are the angles between the principal axes and the edges of the cube before deformation. To be strictly correct, we should have used the angles after deformation, which are slightly different. However, for small strains the errors caused by this simplification are negligible.

The y direction of the cube is one of the principal directions of strain along which the principal stress is zero. Thus, the strain in the y direction $= -1{\cdot}30 \times 10^{-3}$ and so edges parallel to the y direction decrease in length by $1{\cdot}30 \times 10^{-3} \, \text{cm}$.

(e) If the principal strains are labelled as in Fig. 4-12(c), then, from Eqn. (2-26), the angle ϕ which a line makes with the ε_1 principal axis after deformation is given by

$$\tan \phi = \frac{1 + \varepsilon_2}{1 + \varepsilon_1} \tan \theta$$

where θ is the angle the line makes with the axis before deformation. The change in angle during deformation is given by $\tan(\phi - \theta)$. Expanding and substituting from above gives

$$\tan(\phi - \theta) = \frac{(\varepsilon_2 - \varepsilon_1) \tan \theta}{(1 + \varepsilon_1) + (1 + \varepsilon_2) \tan^2 \theta}$$

ε_1 and ε_2 are both very small, and so is $\phi - \theta$ (the change in angle during deformation). We can therefore simplify the expression to

$$(\phi - \theta) = \frac{(\varepsilon_2 - \varepsilon_1) \tan \theta}{1 + \tan^2 \theta} = \frac{(\varepsilon_2 - \varepsilon_1) \sin 2\theta}{2}$$

The change in angle between the edge OX of the cube and the principal axis is obtained by substituting $\theta = 30°$ in this expression giving

$$\phi - \theta = -2\cdot10 \times 10^{-3} \quad \text{radians}$$

The change in angle between OZ and the principal axis is obtained by substituting $\theta = 120°$ giving

$$\phi - \theta = 2\cdot10 \times 10^{-3} \quad \text{radians}$$

From the signs of these quantities, OX rotates clockwise while OZ rotates anticlockwise. Thus, the angle between OX and OZ increases by $4\cdot20 \times 10^{-3}$ radians.

Since the plane containing the x and z axes also contains the principal axes, the y axis must remain perpendicular to this plane during deformation. Therefore the angles between OX and OY and between OZ and OY are unaltered by the deformation.

(f) From Fig. 4-12(a), since $\sigma_{xz} = \sigma_{zx}$

$$F_1 \cos 30° = F_2 \sin 30°$$

(the faces of the cube are of unit area). Therefore

$$F_1/F_2 = \tan 30°$$

We are given that $F_1 + F_2 = 4 \times 10^9$

Solving these two equations for F_1 and F_2 gives

$$F_1 = 1\cdot46 \times 10^9 \text{ dyn} \qquad \text{and} \qquad F_2 = 2\cdot54 \times 10^9 \text{ dyn}$$

Relevant exercises: Nos. 4-3 to 4-6.

4-14 Summary

We have derived the following conclusions about the effects of stress on an isotropically deformable body. These conclusions are only valid for small strain and we have to assume that the properties of the material are such that (a) stress and strain are linearly related, and (b) strain is completely determined by stress.

(i) A dilatational stress produces a dilatational strain (Section 4-3).

(ii) A pure shear stress produces a pure shear strain on principal axes coincident with those of the stress (Section 4-4).

(iii) A deviatoric stress produces a deviatoric strain on principal axes coincident with those of the stress (Section 4-5).

(iv) Any state of stress produces a state of strain on principal axes coincident with those of the stress, and related to it by two material-dependent relationships (Section 4-6).

(v) (a) The constant of proportionality relating dilatational stress to dilatational strain is the bulk modulus, k.

(b) The constant relating the shear stress to the angle of shear is the shear modulus, μ.

(vi) Principal stresses and strains may be related as described in Section 4-11; non-principal stresses and strains are described in Section 4-12. In these calculations it is convenient to use moduli known as the Lamé constants, λ and μ, where $\lambda = k - \frac{2}{3}\mu$ and μ is the shear modulus defined above.

EXERCISES

Sections 4-1 to 4-10

4-1 A rectangular parallelepiped is made from material whose shear and bulk moduli are 3×10^{11} dyn/cm^2 and 10^{12} dyn/cm^2 respectively. The three mutually perpendicular edges of this block form a set of cartesian coordinates and the lengths of the edges along these axes are OX = 11 cm, OY = 8 cm, and OZ = 5 cm. Forces F_x, F_y, and F_z are applied, in the directions indicated by the suffixes, to the faces normal to those directions. Determine the change in length (in cm) of the edges for each of the following sets of forces (in 10^8 dyn).

F_x	F_y	F_z
160	110	88
40	55	88
20	55	−132
−70	55	154

4-2 The block of material in the previous example is subjected to forces of 132×10^8 dyn which are applied to the faces normal to the z direction. The force applied to the face whose outward drawn normal is positive acts in the negative x direction and that applied to the other face acts in the positive x direction. Forces acting in the z direction are applied to the faces normal to the x direction and are of such a magnitude and act in such a direction that the block is in both rotational and translational equilibrium. Determine the change in angle between the edges along the x and z directions.

Sections 4-11 to 4-13

4-3 Fill in the gaps in the following table. The stresses and strains are all principal values, the stresses are in units of 10^9 dyn/cm^2, the strains in units of 10^{-3}, and the moduli in units of 10^{12} dyn/cm^2.

	ε_x	ε_y	ε_z	σ_x	σ_y	σ_z	λ	μ	k
(a)	1·5	2·9	−3·4					0·3	1·0
(b)				3·0	−5·0	−4·0	0·9	0·35	
(c)	1·0	2·5	−2·0	4·5	6·0				

4-4 Fill in the gaps in the following table. The units of stress, strain, and moduli are as in the previous problem.

	ε_{xx}	ε_{yy}	ε_{zz}	ε_{xy}	ε_{xz}	ε_{yz}	σ_{xx}	σ_{yy}
(a)	1·0	−2·0	−1·5	2·5	−3·5	2·0		
(b)				−1·5	2·0	−1·0	−4·0	2·5
(c)	1·2			5·0			1·5	−2·0

	σ_{zz}	σ_{xy}	σ_{xz}	σ_{yz}	λ	μ	k
(a)					1·0	0·5	
(b)	1·0				0·9		1·2
(c)	2·5	2·0	−3·0	−1·0			

4-5 If, in the previous problem, the body is a rectangular parallelepiped of which three mutually perpendicular edges define the coordinate system, determine in each case the forces acting on the faces of the body, the changes in length of its edges, and the changes in the angles between these edges when they have the following lengths:

(a) OX = 3 cm   OY = 10 cm   OZ = 2 cm;
(b) OX = 5 cm   OY = 4 cm   OZ = 8 cm.

4-6 A vertical glass tube is closed at the lower end and open at the other. It is marked with a graduated length scale from its closed end and contains water up to a certain graduation mark. A weight is hung from the lower end of the tube. Show that, for small strain, the change in length of the water column as read from the graduated scale is

$$\frac{l\lambda(3\lambda + 4\mu)\sigma^2}{4\mu^2(3\lambda + 2\mu)^2}$$

where l is the original length, σ the axial stress, and λ and μ are the Lamé constants.

5

Analysis of special types of deformation

5-1 Introduction

In the preceding chapters we have discussed the basic concepts which are necessary for the study of the deformation of solid bodies. One of the purposes of such a study is to enable the changes in the dimensions of a body caused by forces acting on it to be calculated. In this chapter, we consider some problems of practical interest in which the derivation of the state of stress from the applied forces, or the state of strain from the given deformation, is much more difficult than in the problems we have considered up to now. In fact, in some cases a complete analysis is too difficult, and we have to simplify the problem either by making assumptions or by performing experiments. We also meet problems involving non-uniform states of stress and strain.

In many practical cases, the forces applied are contact forces. Often they are exerted by weights which rest on the body, or are attached to it by clamps. In more complicated structures the forces are transmitted by rivets or bolts. It is obvious that the distribution of stress in the neighbourhood of such a point of application is very complicated, much more so than that from the distributed loads we have been considering. This difficulty is general to all problems and so we will deal with it first. Fortunately, it can often be avoided by the use of a principle first stated by Barré de Saint-Venant, a French mathematician, in 1855.

5-2 Saint-Venant's principle

If the distribution of forces acting on a sector of the surface of a body is changed so that the resultant force and couple remain the same, then there will be no change in the strain produced at a sufficient distance from the sector, provided that its area is small compared with the total surface area of the body.

This means that, throughout the bulk of a body, the net effect of a force is independent of the method by which it is applied. It can therefore be replaced by a force having the same resultant magnitude, but applied in such a way that the stress distribution is easily determined. This will not alter the strain throughout most of the body, provided the area on which it is applied is small compared with the total area of the surface of the body.

We can demonstrate this principle by a simple experiment (Fig. 5-1). A rectangular piece of rubber, marked with a grid of small squares, is suspended vertically, and a weight is attached. Any change in size and shape of the squares indicates the state of strain at different places in the sheet. Two

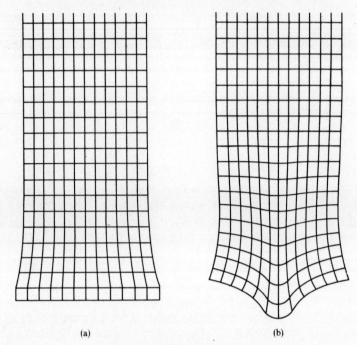

(a)   (b)

Fig. 5-1   Saint-Venant's principle

different methods are used to attach the weight to the sheet. In Fig. 5-1(a) it is distributed uniformly over the lower edge of the sheet by a rigid clamp; in Fig. 5-1(b) it is applied at a point near the lower edge. Comparing Figs. 5-1(a) and 5-1(b), we see that throughout most of the sheet the strain is the same in both experiments. It only changes near the point of application of the load, thus confirming Saint-Venant's principle in this particular case.

In stating Saint-Venant's principle, we use such terms as 'sufficient distance' and 'small' area. We cannot specify these terms more precisely. The principle is an approximation which has never been proved for the general case; it has only been possible to prove it for a few specific examples. Its justification lies in the fact that its application leads to results that can be confirmed by experiment. Thus, as in all approximations, discretion must be used in deciding whether or not it is applicable to a particular problem. The examples in this chapter indicate the sort of situation in which it can be used and the restrictions caused by its use.

5-3   Simple elongation

5-3-1   Definition of the term

Simple elongation is that deformation produced in a long prismatic rod by forces acting parallel to its axis, the resultant force acting at the centroid of the cross section [Fig. 5-2(a)]. Fig. 5-2(b) shows an arrangement by which it may be demonstrated. To ensure that the forces act parallel to the axis, the rod must be mounted vertically, and to ensure that the resultant force acts at the centroid of the cross section, the centre of gravity of the weights must lie on the line of the axis. This is an extremely simple experiment to set up, and it provides a convenient method for the experimental study of the elastic properties of materials.

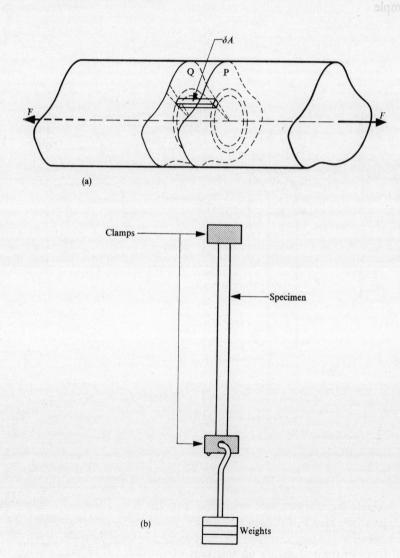

Fig. 5-2 Simple elongation: (a) Type of deformation, (b) Experimental realization

5-3-2 State of stress

Simple elongation is defined in terms of the shape of the specimen and the forces acting on it. The next step is to determine the stress components acting on the faces of the element δA in Fig. 5-2(a). This element is shown enlarged in Fig. 5-3, where the rod is orientated so that its axis coincides with the z axis of a system of cylindrical coordinates.

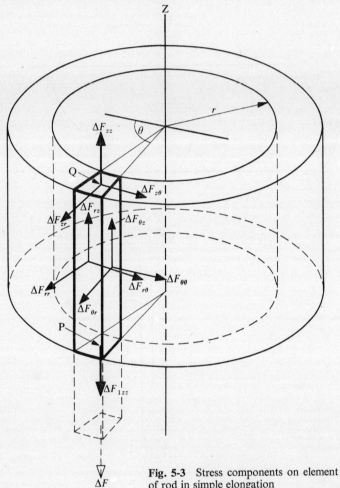

Fig. 5-3 Stress components on element of rod in simple elongation

Suppose the element were separated from the rest of the rod at all points except at the face P, forming a filament. Then the force ΔF_{zz}, which must be applied to the face Q of this filament to maintain equilibrium, is the internal force in the rod acting on this part of the face. If we assume for the moment that the internal forces ΔF_{rz} and $\Delta F_{\theta z}$ are zero, then ΔF_{zz} is equal to ΔF, which is that part of the total applied force that acts on the lower end of the filament. The direction of ΔF is under our control, so let us apply it parallel to the z axis. Then, if

$$\Delta F_{rz} = \Delta F_{\theta z} = 0$$
$$\Delta F_{zz} = \Delta F$$

If the element is now separated from the rest of the filament at P, then, for equilibrium,

$$\Delta F_{1zz} = \Delta F_{zz}$$

Internal forces ΔF_{rz} and $\Delta F_{\theta z}$, assumed above to be zero, could be generated by contact forces applied to the side faces of the rod. However, we know from our definition of simple elongation that no such forces are applied. ΔF_{rz} and $\Delta F_{\theta z}$ could also arise from a *non-uniform strain distribution throughout the rod*. Suppose a neighbouring element were strained differently. Adjacent faces would try to change their dimensions differently, but since the material is continuous this would not be possible, and so internal forces such as ΔF_{rz} would be generated. Thus, the condition for ΔF_{rz}, $\Delta F_{\theta z}$, and $\Delta F_{\theta r}$ to be zero is that the strain must be uniform in all elements. Hence the applied load must be distributed so that ΔF_{zz} produces the same axial stress in all elements.

The axial stress will cause the transverse dimensions of the element to change and this can produce two effects. First, if the change is constrained by the clamp which attaches the load to the rod, the internal force ΔF_{zr} acting on one end face of the element will be different from that on the other end face; and second, if the changed dimensions of the element require more or less space than that allocated by surrounding material, forces ΔF_{rr} and $\Delta F_{\theta\theta}$ will be generated.

Thus, because of the first of these effects, the method of attaching the load to the rod must not constrain dimensional changes, otherwise the stress (and hence the strain) will be non-uniform. Because of the second effect, we need to ensure that the change in transverse dimensions of a free element under stress is the same as the change in the space allocated to it by surrounding material. This is a necessary condition for $\Delta F_{rr} = \Delta F_{\theta\theta} = 0$.

Consider the annulus of radius r shown in Fig. 5-3. Let the axial stress cause a transverse strain of $-\varepsilon$ in an isolated filament lying in this annulus. Since this annulus is made up of a large number of such filaments, all identical, this would cause its circumference to change to $2\pi r(1 - \varepsilon)$, and this, if it is not constrained, would in turn cause its radius to change to $r(1 - \varepsilon)$. Now consider the rod of radius r threading through the annulus. Since strain is uniform, the lateral strain in this rod would be $-\varepsilon$ and its radius would change to $r(1 - \varepsilon)$. Thus it will allow the annulus to assume the radius it requires without constraint, and so ΔF_{rr} and $\Delta F_{\theta\theta}$ will be zero.

So we can say that, provided (a) the force applied to any elemental filament is parallel to the axis of this filament; (b) the applied force is distributed over the end face so as to produce the same axial stress in all filaments; and (c) the method of force application does not constrain the dimensional changes, then all internal forces except ΔF_{zz} are zero. Furthermore, since stress is uniform, the axial stress σ_{zz} is given by the total applied force divided by the total cross-sectional area.

But this conclusion appears to be valid only if the force is applied by certain carefully specified methods. However, it follows from Saint-Venant's principle that the strain (and hence the stress) is independent of the method of force application, except in the region near to where the forces are applied (that is, near the ends of the rod), provided that the area to which these forces are applied is small compared with the total surface area of the rod. This condition will be satisfied if the rod is long and thin. Furthermore, the deformation near the ends of such a rod will contribute negligibly to its total change in dimensions. Thus, for a long thin rod we can calculate the deformation assuming uniform strain (and stress) throughout, whatever the method of force application.

Since the only stress components are those parallel to the axis of the rod, and these are normal stresses, then this axis, the z axis in Fig. 5-3, is a principal axis of stress, and the principal stress along it is F/A, where A is the cross-sectional area. The other principal stresses are zero, and can be directed along any two mutually perpendicular axes x and y in the plane normal to the z axis.

5-3-3 State of strain

If $\sigma = F/A$, the principal stresses are $\sigma_z = \sigma$, $\sigma_x = \sigma_y = 0$. The principal strains can be obtained by substituting these values in Eqns. (4-17):

$$\varepsilon_z = \sigma \frac{(\lambda + \mu)}{\mu(3\lambda + 2\mu)} \tag{5-1a}$$

$$\varepsilon_x = \varepsilon_y = -\sigma \frac{\lambda}{2\mu(3\lambda + 2\mu)} \tag{5-1b}$$

These are the principal strains in the element shown in Fig. 5-3. However, since the strain is uniform, ε_z is also the strain along the axis of the rod, and ε_x the strain in any transverse direction.

5-3-4 Change in dimensions of the rod

From Eqns. (5-1), the applied force will cause the length of the rod to increase and its transverse dimensions to decrease. If l is the original length of the rod, the strained length is $l(1 + \varepsilon_z)$. From Eqn. (5-1b), the strain is the same in all directions in a plane normal to the axis. Thus, if the cross section of the rod is a circle of radius r before deformation, it will be a circle of radius

$r(1 + \varepsilon_x)$ afterwards. The change in cross sections of different shape can be calculated similarly. Since the cross-sectional area of the rod decreases on straining, the value in the strained state should be used in calculating the stress. However, since strains are very small, negligible error is introduced by using the unstrained value.

5-3-5 Young's modulus and Poisson's ratio

From Eqn. (5-1a), the axial strain is given by the product of the axial stress and an expression containing λ and μ. Since λ and μ are both constants, the strain is proportional to the stress. The elastic moduli defined in Section 4-8 were ratios of stress to strain for particular types of deformation. We can now define a further elastic modulus as the ratio of axial stress to axial strain in simple elongation. This is *Young's modulus* (E); it is not independent of the other moduli but can be expressed in terms of them. Thus, from Eqn. (5-1a), E is given by

$$E = \frac{\mu(3\lambda + 2\mu)}{\lambda + \mu}$$

Substituting for λ in terms of k and μ from Eqn. (4-15) gives

$$\frac{1}{E} = \frac{1}{3\mu} + \frac{1}{9k} \tag{5-2}$$

For all materials the bulk modulus is found to be greater than the shear modulus and for most it is more than three times as great. Thus, $1/9k$ is small compared with $1/3\mu$, and so, from Eqn. (5-2), E is determined primarily by the value of μ and, very approximately,

$$E \simeq 3\mu$$

Hence, the deformation must be primarily deviatoric, although there is a small dilatational component.

From Eqn. (5-1b), the transverse strain is also proportional to the axial stress, and so we could define a further elastic constant as the ratio of these two quantities. However, the usual practice is to define this second constant as the ratio of the transverse contraction to the axial extension. This is not, therefore, a modulus, but a dimensionless ratio, called *Poisson's ratio* and denoted by the symbol ν.

So we can write the above definition as

$$\nu = -\varepsilon_x/\varepsilon_z$$

whence, substituting from Eqns. (5-1a) and (5-1b)

$$\nu = \frac{\lambda}{2(\lambda + \mu)}$$

Again, λ can be expressed in terms of k and μ from Eqn. (4-15), giving

$$v = \frac{3k - 2\mu}{6k + 2\mu} \tag{5-3}$$

By rearranging this equation we can see that the value of v must lie between certain limits.

$$k = \frac{2\mu(1 + v)}{3(1 - 2v)} \tag{5-4}$$

From the way in which they are defined, μ and k must both be positive. Equation (5-4) can only be satisfied with positive values of μ and k if v has a value lying between -1 and $\frac{1}{2}$. Substituting $v = \frac{1}{2}$ gives $k/\mu = \infty$, or $k \gg \mu$. Substituting $v = -1$ gives $k/\mu = 0$, or $k \ll \mu$. Thus Poisson's ratio is a measure of the relative magnitude of k and μ. For a material for which v is nearly $\frac{1}{2}$ (e.g., natural rubber), since the bulk modulus must be very much greater than the shear modulus, the dilatational strain in simple elongation must be negligible compared with the deviatoric. For such a material we can assume that deformation takes place at constant volume.

We have shown, then, that in simple elongation the axial strain is proportional both to the axial stress, and to the transverse contraction, and we have defined two new elastic constants from the constants of proportionality. Since simple elongation can quite easily be realized experimentally, these constants can be directly measured, and the other moduli calculated using Eqns. (5-2) and (5-4). The conclusions that both axial and transverse strain are proportional to the axial stress were obtained using Eqns. (4-17), and are therefore subject to the limitations and involve the assumptions introduced in the derivation of these equations, enabling the validity of these assumptions to be investigated experimentally.

5-3-6 Elastic energy stored in simple elongation

In simple elongation a force is applied to the end of a rod. As the rod extends, the point of application of this force moves, and so the force does work on the rod. If the force is reduced the rod will contract, doing work on the force, so this energy must be stored in the deformed rod, and is known as the *stored elastic energy*.

If, in the unstrained state, the cross-sectional area of the rod is A, the force applied to its end is σA (if strains are small). Let the stress be increased to $\sigma + d\sigma$, the strain increasing by $d\varepsilon$. Then, if the unstrained length is l, the work done in this change is $\sigma Al \, d\varepsilon$.

Thus, at strain ε, the stored elastic energy is given by

$$U = Al \int_0^\varepsilon \sigma \, d\varepsilon$$

This equation may be integrated by making the substitution $\sigma = \varepsilon E$, giving

$$U = \tfrac{1}{2}AlE\varepsilon^2$$

Since Al is the volume of the rod, the stored elastic energy per unit volume, u, is given by

$$u = \tfrac{1}{2}E\varepsilon^2 \qquad (5\text{-}5\text{a})$$

or

$$u = \tfrac{1}{2}\sigma\varepsilon \qquad (5\text{-}5\text{b})$$

or

$$u = \frac{\sigma^2}{2E} \qquad (5\text{-}5\text{c})$$

5-3-7 Analysis of simple elongation into dilatational and deviatoric components

Although we have now expressed the change in dimensions of a rod in simple elongation in terms of the force acting and the elastic moduli, it is instructive to analyse the deformation into dilatational and deviatoric components. This will give us an idea of the relative movement of particles of the material during deformation.

Let the longitudinal principal strain be ε_z and the transverse principal strains (which are the same in both the x and y directions) be ε_x. The dilatational strain is given by

$$\varDelta = \varepsilon_z + 2\varepsilon_x$$

Substitution from Eqns. (5-1) gives

$$\varDelta = \frac{\sigma}{3\lambda + 2\mu} \qquad (5\text{-}6)$$

λ and μ are both positive, and so for an extensive stress the volume of the specimen increases. The separation between particles of material will, therefore, increase due to this strain component (Fig. 2-16).

The principal strain of the dilatational component is $\tfrac{1}{3}\varDelta$, and so we can find the principal values of the deviatoric strain component by subtracting this from each of the total principal strains:

$$\varepsilon_{xd} = \varepsilon_{yd} = -\tfrac{1}{3}(\varepsilon_z - \varepsilon_x)$$
$$\varepsilon_{zd} = \tfrac{2}{3}(\varepsilon_z - \varepsilon_x)$$

Substitution from Eqns. (5-1) gives

$$\varepsilon_{xd} = \varepsilon_{yd} = -\sigma/6\mu$$
$$\varepsilon_{zd} = \sigma/3\mu$$

This deviatoric strain and its analysis into two pure shears is shown in Fig. 5-4. It will increase the length of the axis of the rod and decrease its radius, the volume remaining constant. We have already seen that the radius of a rod decreases in simple elongation, so the decrease caused by the deviatoric component must be greater than the increase caused by the dilatational component of the strain.

The relative displacement of particles of material of a body in pure shear is shown in Fig. 2-17. In this figure, the lines of particles which slide are at 45° to the principal axes. From Fig. 5-4, the deviatoric component of simple elongation can be resolved into two equal pure shear strains, the direction

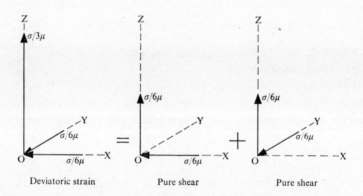

Fig. 5-4 Analysis of deviatoric component of simple elongation into pure shears

of the positive principal strain of each shear coinciding with the axis of the rod. Displacement by the sliding of planes (Fig. 2-17) will occur for both of these shears, and the displacement of any particle will be given by the vector sum of both components. Thus, the deviatoric component of the deformation can be achieved by sliding planes of particles along lines lying in their planes and at 45° to the rod axis.

5-3-8 Worked example

If the bulk and shear moduli of rubber are $1 \cdot 0 \times 10^{11}$ dyn/cm² and $1 \cdot 0 \times 10^7$ dyn/cm², respectively, calculate the value of Poisson's ratio. Determine the dilatational strain as a fraction of the longitudinal strain in simple elongation and deduce an approximation that can be made in calculations of the extension of rubber.

Rubber can easily be stretched to strains greater than those to which small strain elasticity theory can be applied. Use the approximation just deduced to determine the strain at which a value of Poisson's ratio, determined by measuring the longitudinal and lateral strain, would begin to differ from that calculated above.

From Eqn. (5-3), Poisson's ratio is given by

$$\nu = \frac{3k - 2\mu}{6k + 2\mu}$$

Substituting the values given for k and μ gives

$$\nu = \frac{30,000 - 2 \cdot 0}{60,000 + 2 \cdot 0}$$

From the accuracy with which the moduli are given, the $2 \cdot 0$ can be neglected, and

$$\nu = 0 \cdot 50$$

If σ is the longitudinal stress, then, by Eqn. (5-6), the dilatational strain is given by

$$\Delta = \frac{\sigma}{3\lambda + 2\mu}$$

The longitudinal strain, ε, is given by Eqn. (5-1a) and is

$$\varepsilon = \frac{\sigma(\lambda + \mu)}{\mu(3\lambda + 2\mu)}$$

Therefore,

$$\frac{\Delta}{\varepsilon} = \frac{\mu}{\lambda + \mu}$$

Since, by Eqn. (4-15), $\lambda = k - \frac{2}{3}\mu$, then

$$\frac{\Delta}{\varepsilon} = \frac{3\mu}{3k + \mu}$$

Substituting the values given for k and μ gives

$$\frac{\Delta}{\varepsilon} = 10^{-5}$$

Hence, the dilatational strain is negligible compared with the longitudinal strain and can be neglected in simple elongation. That is, extension takes place at approximately constant volume.

Using this approximation we can determine the lateral strain for any value of the longitudinal strain. Consider a rod of length l and radius r. Let the longitudinal strain be ε_z and the lateral ε_x. Then, since extension takes place at constant volume,

$$\pi r^2 l = \pi r^2 (1 + \varepsilon_x)^2 l(1 + \varepsilon_z)$$

Expanding this equation by the binomial theorem gives

$$\varepsilon_x = -\varepsilon_z/2 + 3\varepsilon_z^2/8 + \cdots$$

Therefore, since $\quad v = -\varepsilon_x/\varepsilon_z$

$$v = 1/2 - 3\varepsilon_z/8$$

The value at which v will begin to diverge from $\frac{1}{2}$ depends on the accuracy of measurement. If, for example, a difference of 0.005 in v can be detected, then $3\varepsilon_z/8 > 0.005$ for v to differ significantly from $\frac{1}{2}$. That is, v will begin to differ from 0.5 when $\varepsilon_z > 0.013$ (i.e., 1.3%).

5-4 Twisting of shafts

5-4-1 Definition of term

Twisting is the term used to describe the deformation which occurs whenever a couple acts in the plane of the cross section of a shaft. This happens in

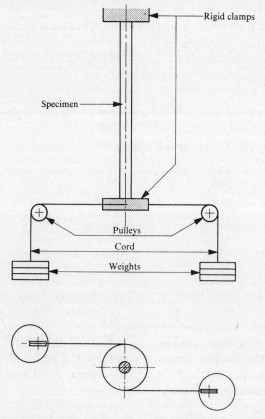

Fig. 5-5 Experimental method of applying torque to rod

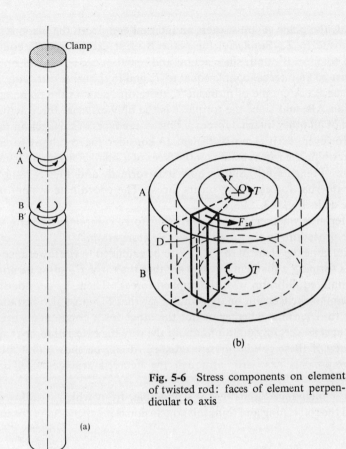

(a)

(b)

Fig. 5-6 Stress components on element of twisted rod: faces of element perpendicular to axis

any rod, such as an axle or propeller shaft, which is used to transmit torque. It can be demonstrated by the apparatus shown in Fig. 5-5. The torque is applied so that the couple in the plane of the cross section is the only force acting. The weight of the lower clamp must not cause an axial tension in the rod, neither do the weights produce a couple in the vertical plane. We shall consider a rod of circular cross section, since this is the only case which can be solved simply.

5-4-2 State of stress

The rod and the applied couple are shown schematically in Fig. 5-6(a). To determine the state of stress, we have to find the internal forces acting on the faces of a typical element of the rod. Suppose the rod is cut through at the cross section AA′; the lower part of the rod would spin under the action of the externally applied couple T. To prevent this spin occurring before the rod is cut, the plane A′ must exert an internal couple on the plane A, equal and opposite to T. Similarly, the plane B must exert a couple equal and opposite to T on B′, and, since action and reaction are equal and opposite, there must be an internal couple equal to T, and in the same direction, acting on the face B. A couple of moment T, therefore, acts on the plane faces of the section AB, and since the torque T is the only external force acting, the resultant of all other internal forces acting on the faces of this section must be zero. However, as it is not sufficient to consider the resultant forces, we must determine the forces acting on the faces of a small element cut from such a section. Figure 5-6(b) shows how it is formed, and Fig. 5-7 shows all possible internal forces acting on its faces. The coordinate system used is the same as in Fig. 5-3.

To determine whether or not a particular force component is zero, we use a similar procedure to that used for simple elongation.

If we imagine the element in Fig. 5-7 to be extended to the lower face of the rod, thus forming a filament, we can see that the force $F_{z\theta}$ must be non-zero (to maintain equilibrium with the applied force). Then, by considering the equilibrium of the element on its own, we see that $F_{\theta z}$ must also be non-zero. In order to determine whether or not the other force components are zero, we must specify certain conditions about the way the external force is applied to the rod. If these conditions are satisfied, it can be shown that all other force components are zero, although the detailed argument will not be given here.

We now introduce Saint-Venant's principle, from which it follows that, provided the rod is long and thin, this conclusion is independent of the method

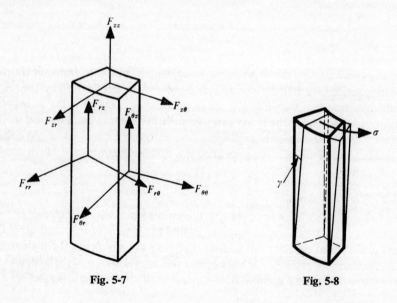

Fig. 5-7   Fig. 5-8

of load application, except for small regions near the ends of the rod. Since these contribute negligibly to the total change in dimensions, they can be ignored.

Thus, the only non-zero forces acting on the faces of the element are those shown in Fig. 5-8; these clearly constitute a shearing system. The forces $F_{z\theta}$ arise as a direct consequence of the application of the external torque. External forces do not, however, *directly* cause $F_{\theta z}$. The forces $F_{z\theta}$ constitute a couple which tends to make the element rotate. This rotation deforms surrounding material, thus developing $F_{\theta z}$. From Section 3-15, the element is therefore in simple shear.

For the moment, we can only determine the type of stress acting on the element, not its magnitude. This is because the stress is non-uniformly distributed over the cross section. However, we can leave it as an unknown and express the deformation of an element in terms of it. Integration over the cross section will relate the total torque, which we do know, to the net effect of the deformation of all elements, which we want to find.

5-4-3 State of strain

Since the stress is a simple shear, the strain must also be a simple shear (Section 4-9). The element will therefore deform as in Figs. 5-9 and 5-10. Furthermore, if σ is the shear stress and γ the shear strain, then

$$\sigma = \mu\gamma$$

where μ is the shear modulus of elasticity. However, since σ is not known, we cannot evaluate the strain at this stage.

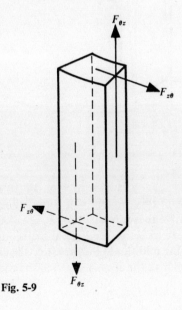

Fig. 5-9

5-4-4 Deformation of the rod

We determine the deformation of the rod from the total effect of the strain in all the elements similar to that in Fig. 5-9. Before doing this in detail, we can state the following general conclusions about the change in dimensions of the rod.

Length of rod From Section 2-16 we see that in simple shear strain the perpendicular distance between planes such as A and B in Fig. 5-6(b) remains unaltered. It therefore follows that the length of the rod is unchanged during the deformation.

Radius of the rod The volume of a body in simple shear is unchanged by the deformation. Hence, since the length of the rod is unaltered, its radius cannot change.

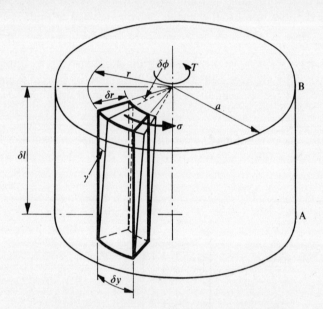

Fig. 5-10 Deformation of element of twisted rod

If all elements in the section shown in Fig. 5-6(b) deform as shown in Fig. 5-10, then one face of this element will rotate about the axis relative to the other. It is this relative rotation which must be calculated in terms of the applied torque, the shear modulus, and the dimensions of the rod. To do this we need the following additional information about the deformation.

(i) Planes perpendicular to the cross section of the rod before deformation will be plane and perpendicular afterwards. This is because the distance between planes such as A and B in Fig. 5-6(b) is constant in simple shear, whatever the magnitude of the strain. Thus, even if different elements in this section shear by different amounts their perpendicular heights will all be the same.

(ii) Straight radial lines drawn in the section AB before deformation will be straight after deformation. The lines would only curve as a result of shear in the plane of the cross section of the rod. To produce this, forces $F_{r\theta}$ in Fig. 5-7 would have to act, but we have already stated that these are zero.

(iii) The axis of the rod is straight after deformation. If the axis were curved, then the edges of the element would be curved. Simple shear does not produce this curvature.

We can now analyse Fig. 5-10, which shows the section AB before and after deformation. In this figure, the lower face of the element is shown occupying the same position in space before and after deformation. Actually, from Fig. 5-6(a), only the upper surface of the element in contact with the clamp would have this property; any other element would be moved as a rigid body round the annulus. We have, in fact, superimposed a rigid body rotation in Fig. 5-10, but this will not affect our analysis, since we are only interested in the relative rotation between faces. From Fig. 5-10,

$$r \, \delta\phi = \gamma \, \delta l$$

and, since
$$\sigma = \mu\gamma$$

$$\sigma = \mu r \frac{\delta\phi}{\delta l} \tag{5-7}$$

σ is the stress acting on the upper surface of the element and so the force on this surface is $\sigma \, \delta r \, \delta y$, and the couple acting about the centre of the rod due to this force is $\sigma r \, \delta r \, \delta y$. Therefore, from Eqn. (5-7), the couple about the axis of the rod due to the force on the element is

$$\mu r^2 \frac{\delta\phi}{\delta l} \, \delta r \, \delta y$$

If we integrate this expression over the whole surface of the cross section, it will be equal to the applied torque. First, however, we must integrate round the annulus of which the element is a part. Since cylindrical symmetry exists, the only variable at this stage is δy; the integral of this round the annulus is $2\pi r$. Thus, the couple about the axis of the rod due to the force on the annulus is

$$2\pi\mu r^3 \frac{\delta\phi}{\delta l} \, \delta r$$

We can now integrate this expression between $r = 0$ and $r = a$ to obtain the couple about the axis of the rod due to the force on the entire cross section, and this is equal to the applied torque T. Thus

$$T = 2\pi\mu \frac{\delta\phi}{\delta l} \int_0^a r^3 \, dr \tag{5-8}$$

From conditions (i) and (ii) above, $\delta\phi$ and δl are both independent of r.

Thus, $$T = \frac{\pi\mu a^4}{2} \frac{\delta\phi}{\delta l} \tag{5-9}$$

Since longitudinal symmetry exists, ϕ and l are the only variables along the length of the rod, so we can integrate this expression to give the relative rotation of the ends of a rod of length l as

$$\phi = \frac{2Tl}{\pi\mu a^4} \tag{5-10a}$$

which can also be written as

$$\frac{T}{\phi} = \frac{\pi\mu a^4}{2l} \tag{5-10b}$$

The quantity T/ϕ, which is the torque required to produce unit twist in the rod, is known as the torsional rigidity of the rod. It is a quantity which is easily determined experimentally, and which, for rods of circular cross section, can be used with Eqn. (5-10) to determine the shear modulus.

If, instead of a solid rod, a tube of outside and inside radii a_2 and a_1, respectively, is subjected to a torque T, the torsional rigidity can be found by integrating Eqn. (5-8) between a_1 and a_2, giving

$$\frac{T}{\phi} = \frac{\pi\mu(a_2^4 - a_1^4)}{2l} \tag{5-11}$$

5-4-5   Distribution of stress and strain

We have already stated that cylindrical symmetry exists, so that at all points on the annulus (Fig. 5-10) the stress must be the same. Since longitudinal symmetry also exists, the stress is the same at whatever point along the axis the annulus is situated. But we have not yet determined the magnitude of this stress.

We do this by eliminating $\delta\phi/\delta l$ between Eqns. (5-7) and (5-9):

$$\sigma = \frac{2Tr}{\pi a^4} \tag{5-12}$$

Thus the stress is not uniform but will increase from zero at the centre of the rod to a maximum value of $2T/\pi a^3$ at the circumference.

We determine the strain by substituting $\mu\gamma$ for σ in Eqn. (5-12), giving the angle of shear as

$$\gamma = \frac{2Tr}{\pi\mu a^4}$$

5-4-6 Elastic energy stored in twisting

If an increase of dT in the applied torque causes the relative rotation of the ends of the rod to increase by $d\phi$, then the work done on the rod in twisting it through an angle ϕ is

$$\int_0^\phi T \, d\phi$$

Thus, from Eqn. (5-10a), the stored elastic energy is

$$U = \frac{\pi\mu a^4 \phi^2}{4l} \tag{5-13a}$$

$$U = \frac{lT^2}{\pi\mu a^4} \tag{5-13b}$$

5-4-7 Worked examples

1. A tubular shaft of fixed length is to be designed so that, when it is subjected to a torque T, the maximum angle of shear in the material is γ. Show that, whatever value is chosen for the outer radius of this shaft, a reduction in mass can be achieved, without increasing γ, by further increasing this radius.

If practical considerations limited the radius which could be used, but it was important to keep mass to a minimum, what combination of physical properties would determine your choice of material?

From Eqn. (5-11),

$$T = \frac{\pi\mu\phi}{2l}(a_2^4 - a_1^4)$$

From Section 5-4-5, the angle of shear has its maximum value γ at the outer radius, and from Fig. 5-10 this is given by

$$\gamma = \phi a_2/l \tag{5-14}$$

If the mass of the shaft is M and the density of the material ρ, then

$$M = \pi(a_2^2 - a_1^2)l\rho \tag{5-15}$$

Substituting from Eqns. (5-14) and (5-15) in (5-11) gives

$$\frac{Tl}{\gamma} = \frac{M\mu}{2\rho a_2}\left(2a_2^2 - \frac{M}{\pi l\rho}\right)$$

This equation expresses the relationship between M and a_2. Under the conditions stated in the problem, all other terms are constants. The equation may be simplified by replacing groups of constants according to the equations

$$K_1 = T/\pi\mu\gamma$$
$$K_2 = \pi l\rho$$

giving
$$K_1 K_2 = 2Ma_2 - \frac{M^2}{2K_2 a_2} \qquad (5\text{-}16)$$

which, solved for M, gives

$$M = 2K_2 a_2^2 \left[1 \pm \left(1 - \frac{K_1}{2a_2^3}\right)^{1/2}\right] \qquad (5\text{-}17)$$

We have to determine now whether the positive or negative root is correct.

Assume, first of all, that the positive root is the correct one, so that Eqn. (5-17) becomes

$$M = 2K_2 a_2^2 \left[1 + \left(1 - \frac{K_1}{2a_2^3}\right)^{1/2}\right]$$

Since the term in the square brackets must be greater than unity, it follows that

$$M > 2K_2 a_2^2$$

or, since
$$K_2 = \pi l\rho$$

$$M > 2\pi\rho l a_2^2$$

if the positive root is correct. However, from Eqn. (5-15),

$$M = \pi\rho l(a_2^2 - a_1^2)$$

which means that
$$M < \pi\rho l a_2^2$$

Therefore the positive root leads to an incorrect statement, and we must take the negative one:

$$M = 2K_2 a_2^2 \left[1 - \left(1 - \frac{K_1}{2a_2^3}\right)^{1/2}\right]$$

Since $(1 - K_1/2a_2^3)^{1/2}$ must be real, $K_1/2a_2^3 < 1$, and a binomial expansion can be used to simplify this equation, giving

$$M = K_1 K_2 \left[\frac{1}{2a_2} + \frac{K_1}{16a_2^4} + \text{further positive terms with } a_2 \text{ in the denominator}\right]$$

Therefore, whatever the value of a_2, an increase in its magnitude will reduce the value of the terms in the square bracket, and hence will cause M to decrease.

Now
$$M = K_1 K_2/2a_2 + \text{smaller terms}$$

In the second part of the problem, choice of material is allowed, i.e., μ and ρ, and hence K_1 and K_2, may vary, but the outside radius of the rod, a_2, is fixed. Thus the shaft with the smallest mass will be that made from the material which has the smallest value of $K_1 K_2$. Now

$$K_1 K_2 = {}_{\blacksquare}^{r} Tl\rho/\mu\gamma$$

and in this expression only ρ and μ depend upon the material, l being fixed. Hence, the material having the smallest value of ρ/μ should be chosen.

2. A rod with a cylindrical cross section of radius a is subjected to a torque T. Calculate the stored elastic energy per unit volume at a point in the rod situated at a distance r from the axis.

Consider the small block of material shown in Fig. 5-11 subjected to a

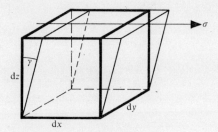

Fig. 5-11 Worked example 5-4-7(2)

shearing stress σ. The force applied to the element is $\sigma\,dx\,dy$, and the distance moved by its point of application is $\gamma\,dz$. Thus, the work done by the force is

$$dx\,dy\,dz \int_0^\gamma \sigma\,d\gamma$$

Thus, the work done per unit volume is $\sigma^2/2\mu$. The material in the rod will be subjected to this state of stress, and, from Eqn. (5-12), the stored elastic energy per unit volume is given by

$$u = \frac{2T^2 r^2}{\mu\pi^2 a^8}$$

5-5   Bending of beams

5-5-1   Definition of the term

In the two previous cases, we have defined the shape of a body and the forces acting on it and, from this information, calculated the deformation. We now adopt the converse procedure. We define the deformation and calculate the loads necessary to produce this deformation.

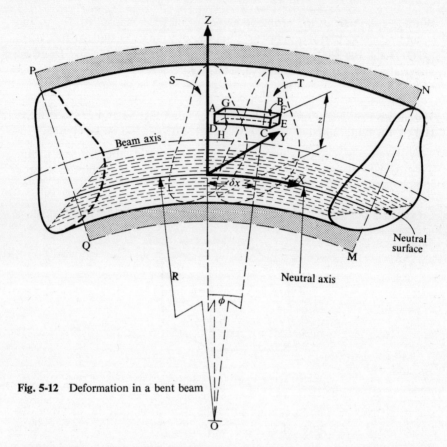

Fig. 5-12   Deformation in a bent beam

The body is again in the shape of a long prismatic rod, but in this case the cross section need not necessarily be circular. The only restriction on its shape is that it is symmetrical about the line MN in Fig. 5-12. The plane PQMN, containing this line and the axis of the rod, is called the *plane of symmetry*.

Before deformation, the axis of the rod is straight; after deformation it is bent so that it still lies in the plane of symmetry and short elements can be considered as arcs of circles. Planes such as S and T, which are normal to the axis before deformation, remain plane and normal to the axis afterwards. Lines of particles such as AB, which are parallel to the axis before deformation, remain parallel afterwards. Cross sections of the rod remain symmetrical about the plane of symmetry after deformation.

5-5-2   State of strain

Consider the element of material shown in Fig. 5-12. The faces ABCD and EFGH are rectangular before deformation and parallel to the plane of symmetry. The faces CEFB and DHGA are also rectangular and are parts of the planes S and T normal to the rod axis. We use the system of coordinates shown in the diagram, in which the z axis lies in the plane of symmetry, and the x axis is parallel to the rod axis.

Since the planes CEFB and DHGA are normal to the rod axis, and this is initially straight, the unstrained lengths of the lines AB and CD are equal. After deformation the element of axis is bent into the arc of a circle. The planes CEFB and DHGA, however, remain planar and normal to the axis, so that lines of particles such as DC and AB become arcs of concentric circles and must change in length by different amounts. The strain ε_{xx}, therefore, varies over the cross section and might be a function of z or y. However, if planes normal to the axis before deformation are plane and normal after, and PQMN is a plane of symmetry of the deformed rod, it follows that ε_{xx} is independent of y, and so it is a function only of z.

Since the plane CEFB remains normal to the rod axis on deformation and lines such as AB remain parallel to the axis, ε_{xy} and ε_{xz} must be zero. Since the rectangle CEFB can change its size and shape without affecting the specification of the deformation given in Section 5-5-1, there is no restriction on the values which can be taken by ε_{yy}, ε_{zz}, and ε_{yz}. For the time being, therefore, we will not specify the values of these strain components.

From the above, the important parameters defining the deformation are ε_{xx} and its variation with z, which we will now determine.

If the axes of symmetry of the planes S and T are produced, they will intersect at O, the centre of curvature of the element. Clearly, the deformed length of AB might be greater or less than the undeformed length, depending upon its coordinate, and there will be one position where these lengths are equal. In other words, *a surface exists in which any line drawn parallel to the axis is unchanged in length on deformation.* This surface is shown shaded in Fig. 5-12 and is known as the *neutral surface*. The line formed by the intersection of the neutral surface with the plane of symmetry is called the *neutral axis,* and this is chosen as the x axis.

Since O must be the centre of curvature of the element δx of the neutral axis, then R is its radius of curvature. Also, δx must be the unstrained length of AB, so the extensional strain of AB is given by

$$\varepsilon_{xx} = \frac{(R + z)\phi - \delta x}{\delta x}$$

Now $$\delta x = R\phi$$

and therefore $$\varepsilon_{xx} = z/R \tag{5-18}$$

The extensional strain in the element shown in Fig. 5-12 is therefore related to the radius of curvature of the beam axis by Eqn. (5-18). Of the other strain components, ε_{xy} and ε_{xz} must be zero, and ε_{yy}, ε_{zz}, and ε_{zy} cannot be determined from the specification of the deformation.

Now if the element were in simple elongation, ε_{xz}, ε_{xy}, and ε_{zy} would all be zero, and ε_{yy} and ε_{zz} would be determined by ε_{xx} and Poisson's ratio of the material. This state of strain would therefore satisfy the above condition, and we shall consider it further.

Elements above the neutral axis, such as that shown in Fig. 5-12, will increase in length, and therefore decrease in cross section, if they are in simple elongation. Similarly, elements below the neutral axis will decrease in length and increase in cross section. Therefore, the lateral dimension of the beam will increase below the neutral surface, and decrease above it (Fig. 5-13). To accommodate this dimensional change, the beam must bend

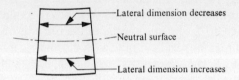

Fig. 5-13   Shape of initially rectangular cross section of bent beam

in a plane perpendicular to the plane of symmetry, the direction of this bending being opposite to that occurring in the plane of symmetry. This effect is known as *anticlastic bending* and is illustrated in Fig. 5-14. However, its effect is negligible in long thin beams such as those we are considering, and we will ignore it.

The transverse strains will have another effect. The quantity z in Eqn. (5-18) is the perpendicular distance of the element from the neutral surface in the strained state. Since ε_{zz} is not zero, this is different from the distance in the unstrained state. However, since strain is small, the difference will be negligible, and will be ignored.

5-5-3   State of stress
Since the element is in simple elongation the only stress acting will be the normal stress σ_{xx}. From Eqn. (5-18)

$$\sigma_{xx} = Ez/R \qquad (5\text{-}19)$$

5-5-4   Forces acting on the beam
Figure 5-15 is a section of Fig. 5-12, the plane of symmetry being the plane of the paper. If the cross-sectional area of the element is δA, the internal force acting on the face AC is $\sigma_{xx}\,\delta A$ directed outwards. A filament of material can be formed by producing the edges of the element, such as AB, to one end of the bar. For the filament to be in equilibrium an external force of

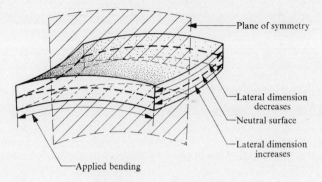

Fig. 5-14 Anticlastic bending

$\sigma_{xx}\,\delta A$ must be applied to this end, and be directed outwards. Had the filament been below the neutral axis, the extensional strain would have been negative and so the internal force, and hence the externally applied force, would have to be directed inwards. We can find the distribution of applied force over the ends of the bar by considering a number of such filaments. From Eqn. (5-19), the force applied to the ends of these filaments must be proportional to their distance from the neutral axis. Hence, the distribution of force required to produce bending as defined in Section 5-5-1 is as illustrated on the left-hand end of the bar in Fig. 5-15.

From Saint-Venant's principle, provided the rod is long and thin, any applied force which has the same resultant as this distribution will produce the same deformation. From Fig. 5-15, the resultant comprises a couple in the plane of symmetry. There might also be an axial resultant force depending upon whether or not the forces above the neutral axis are equal to those below. If such an axial resultant exists, it will affect the length of the bar and not its bending. It is the couple which produces the bending and this couple is called the *bending moment*. We go on now to determine the bending moment in terms of the radius of curvature, the geometry of cross section, and Young's modulus.

Since the extensional strain in an element depends only on its distance from the neutral surface, all elements in the shaded area of Fig. 5-15 will have

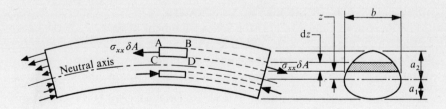

Fig. 5-15 Force distribution to produce bending

the same strain. Hence, from Eqn. (5-19), the external force applied to this shaded area will be

$$\frac{bEz \, dz}{R}$$

and its moment about the neutral axis will be†

$$\frac{bEz^2 \, dz}{R}$$

Therefore, the total axial force applied to the end of the bar will be given by

$$F = \frac{E}{R} \int_{a_1}^{a_2} bz \, dz \qquad (5\text{-}20)$$

and the bending moment will be given by

$$M = \frac{E}{R} \int_{-a_1}^{a_2} bz^2 \, dz \qquad (5\text{-}21)$$

The quantities F and M can be determined from the load applied. If we know this, together with the shape of the beam and the Young's modulus of the material, there are only two unknowns in these equations—the position of the neutral axis and the radius of curvature—which can therefore be determined. The only case we will consider in detail will be that in which the axial force F is zero. This is called *pure bending*.

From Eqn. (5-20), if $F = 0$, since E/R cannot be zero, the integral must be zero. But the equation

$$\int_{-a_1}^{a_2} bz \, dz = 0$$

expresses the condition that the line $z = 0$ contains the centroid of the area of cross section. Now the neutral surface has been taken as the plane for which $z = 0$, and therefore, in pure bending, the neutral surface contains the line of centroids of the cross section, which is, of course, the axis of the beam. Since this axis must lie in the plane of symmetry, and since the neutral axis is formed by the intersection of the neutral surface with the plane of symmetry, it follows that, *in pure bending, the neutral axis coincides with the beam axis*. It can thus be easily located.

The integral in Eqn. (5-21) is the second moment of area, I, of the cross section about the line $z = 0$. This line is normal to the plane of symmetry, and, in the case of pure bending, passes through the beam axis. We can, therefore, calculate I from the geometry of the cross section, and substitution

---

† The quantity $bz^2 \, dz$ is the area of the shaded element multiplied by z^2. The integral of this quantity over the entire cross section is known as the second moment of area of the cross section about the line chosen as the y axis.

in Eqn. (5-21) gives

$$M = EI/R \qquad (5\text{-}22)$$

Thus, if a couple of magnitude M acts in the plane of symmetry of a beam, that beam will be bent into an arc of a circle and the radius of curvature of the neutral axis will be given by Eqn. (5-22).

5-6 Deformation of loaded beams

5-6-1 Internal forces in loaded beams

In Section 5-5 we determined the type of force distribution which must be applied to a beam to produce bending, and we related this to the radius of curvature of the bent beam, the beam geometry, and the Young's modulus of the material. However, in many practical situations the method of loading the beam not only leads to a bending moment but also applies other deforming forces. Furthermore, neither these, nor the bending moment, are necessarily uniform along the beam.

There are many different ways in which loads can be applied to a beam so

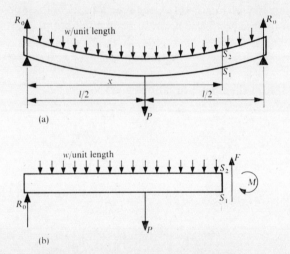

(a)

(b)

Fig. 5-16 Internal forces in loaded beam

as to produce this kind of deformation. Each of these presents its own problems and so we cannot make a general analysis. We shall consider only two typical examples. The method we use to determine the deformation in these examples will enable us to solve a wide range of problems.

We consider first the case shown in Fig. 5-16(a). The beam is held on supports at its ends and a force P is applied at the centre. In addition, the material of the beam has a weight which is distributed along its length, exerting a force w per unit length. We can determine the internal forces acting across the plane at S_1S_2 by imagining the beam to be cut through at this plane and finding the forces which must be applied to the cut face to maintain equilibrium in the left-hand part of the beam. From Fig. 5-16(b), we can see that a couple M (a bending moment) and a vertical force F (a shear force) must be applied to maintain equilibrium. These are the internal forces.

We use the following sign convention for the directions of forces and couples in bending. *The bending moment is positive if it tends to stretch the upper part of the beam and compress the lower. Shear forces are positive if they tend to rotate the section in an anticlockwise direction.* Thus both the shear force and the bending moment have been drawn in the positive direction in Fig. 5-16(b).

For the section to be in equilibrium

$$F = P + wx - R_0$$

$$M = P\left(x - \frac{l}{2}\right) + \frac{wx^2}{2} - R_0x$$

From the symmetry of Fig. 5-16(a), the reactions at both supports must be equal, and so for the complete beam to be in equilibrium

$$R_0 = \frac{P + wl}{2}$$

Substituting this in the above equations gives

$$F = \frac{P}{2} + w\left(x - \frac{l}{2}\right) \tag{5-23a}$$

$$M = (x - l)\left(\frac{P + wx}{2}\right) \tag{5-23b}$$

If the section S_1S_2 had been taken to the left of P, we would have obtained the equations

$$F = -\left[\frac{P}{2} + w\left(\frac{l}{2} - x\right)\right] \tag{5-23c}$$

$$M = -\frac{x}{2}[P + w(l - x)] \tag{5-23d}$$

The graphs of Fig. 5-17, showing the variation of F and M with the distance x along the bar, have been plotted from these four equations.

The distribution of shear force and bending moment can usually be found by methods similar to that used above, but there are two points to be noted.

First, different expressions apply to the portions of the beam on either side of the concentrated load P. In general, a change in the expressions for F and M occurs wherever the force applied to a beam changes discontinuously,

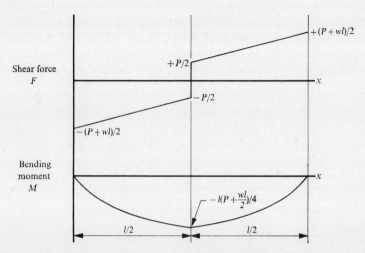

Fig. 5-17   Distribution of shear force and bending moment in loaded beam of Fig. 5-16

as at a point of application of a concentrated load, or at a support. Different expressions must therefore be derived for each section of the beam between such points. If the load is very complicated, this can be inconvenient, and so methods, which are beyond the scope of this book, have been developed to avoid the difficulty. These methods are described in engineering texts on the theory of structures.

The second point is that it is not always possible to determine the magnitudes of all the reactions from the static equilibrium of the entire beam, as we did above. Such problems are called *statically indeterminate* and our second example, shown in Fig. 5-18(a), illustrates this type of problem.

The beam is clamped in a horizontal position at the left-hand end, and passes over a support A which is at the same height as the clamp. The beam has a weight per unit length w, and a force P is applied to the right-hand end. The clamp will hold the beam in a horizontal position at the point where it emerges, even though the other applied forces are tending to bend it. Thus, the clamp must exert a couple M_0 on the beam. It also exerts a reaction force R_1 supporting part of the weight of the beam. To simplify the algebraic expressions we shall derive, let us assume that $l_1 = \frac{2}{3}l$ and $wl = P/10$. Then, for equilibrium,

$$R_1 + R_2 = 11P/10 \tag{5-24a}$$

$$M_0 = l\left(\frac{21P}{20} - \frac{2R_2}{3}\right) \tag{5-24b}$$

These two equations contain three unknown support reactions, R_1, R_2, and M_0.

We can determine the internal forces by imagining the beam cut through

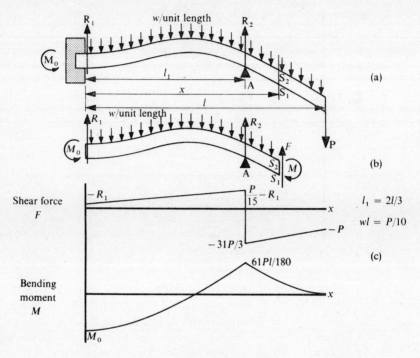

Fig. 5-18

at S_1S_2 and finding the force and couple which must be applied to the cut face to keep the left-hand section in equilibrium. From Fig. 5-18(b), if S_1S_2 lies to the right of A,

$$M = \frac{P(x^2 - 22xl + 21l^2)}{20l} \qquad (5\text{-}25a)$$

$$F = \frac{P(x - 11l)}{10l} \qquad (5\text{-}25b)$$

where R_1, R_2, and M_0 have been eliminated using Eqns. (5-24). Similarly, if S_1S_2 lies to the left of A,

$$M = \frac{P(x^2 - 22xl + 21l^2)}{20l} + \frac{R_2(3x - 2l)}{3} \qquad (5\text{-}25c)$$

$$F = \frac{P(x - 11l)}{10l} + R_2 \qquad (5\text{-}25d)$$

but we cannot eliminate R_2 at this stage. We need first to determine the deformation of the beam, when the condition that its displacement and slope at A must be continuous will allow R_2 to be calculated. Graphs showing the variation of F and M along the beam are shown in Fig. 5-18(c).

5-6-2 Deformation of loaded beam due to bending moment

We see that, in the examples considered above, part of the internal forces acting on the faces of a small element of the beam at S_1S_2 constitutes a couple. This couple will bend the neutral axis of the element into the arc of a circle of a radius given by Eqn. (5-22). Since M varies from point to point along the beam, the radius of curvature will also vary, and will be given, for the part of the beam to the right of P in Fig. 5-16(a), by the equation

$$\frac{EI}{R} = (x - l)\left(\frac{P + wx}{2}\right)$$

However, it is not very helpful to describe the deformation in terms of the radius of curvature of the neutral axis. It is better to express it in terms of the deflection of this axis from the position it occupies in the undeformed state.

In Fig. 5-19, the x axis represents the neutral axis in the undeformed state, and the line OA represents the same axis in the deformed state. The y axis is taken as positive vertically downwards. Thus, if the element MN occupies the position M'N' in the deformed state, we have to derive an equation relating R and y.

If the tangents at M' and N' make angles ψ and $\psi + \mathrm{d}\psi$, respectively, with the x axis, then the angle M'CN' is $\mathrm{d}\psi$. Thus

$$\frac{1}{R} = \frac{\mathrm{d}\psi}{\mathrm{d}s} \tag{5-26}$$

and

$$\tan \psi = \frac{\mathrm{d}y}{\mathrm{d}x} \tag{5-27}$$

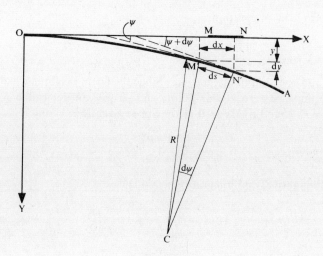

Fig. 5-19

Differentiating Eqn. (5-27) with respect to s, and substituting from Eqn. (5-26), gives

$$\frac{1}{R}\sec^2\psi = \frac{d^2y}{dx^2}\frac{dx}{ds}$$

Since

$$\frac{dx}{ds} = \cos\psi$$

$$\frac{1}{R} = \cos^3\psi\frac{d^2y}{dx^2}$$

Since strains must be small, it follows that ψ must be small. Thus, $\cos\psi$, and hence $\cos^3\psi$, will be very nearly equal to unity. Hence,

$$\frac{1}{R} = \frac{d^2y}{dx^2}$$

and substituting in Eqn. (5-22) gives

$$EI\frac{d^2y}{dx^2} = M \tag{5-28}$$

We can now evaluate the deformation of the beams in the examples we have been considering, starting with the example shown in Fig. 5-16(a). Taking the portion to the left of P in this figure, and substituting from Eqn. (5-28) in (5-23d) gives

$$2EI\frac{d^2y}{dx^2} = -x(P + wl) + wx^2$$

and therefore, by integration,

$$2EI\frac{dy}{dx} = -\frac{x^2}{2}(P + wl) + \frac{wx^3}{3} + K_1$$

where K_1 is a constant of integration. Since the beam is symmetrical about P, the slope must be zero at this point, and so we can determine K_1 by substituting $x = \frac{1}{2}l$ and $dy/dx = 0$ in the above equation, giving

$$K_1 = \frac{l^2}{4}\left(\frac{P}{2} + \frac{wl}{3}\right)$$

Substituting above, and integrating again, gives

$$4EIy = -\frac{x^3}{3}(P + wl) + \frac{wx^4}{6} + \frac{l^2x}{2}\left(\frac{P}{2} + \frac{wl}{3}\right) + K_2$$

The beam is supported at $x = 0$ and at this point $y = 0$, so $K_2 = 0$. Thus, the deflection of the neutral axis of the beam from its undeformed position, due to bending, is given by

$$y = \frac{x}{4EI}\left[\frac{wx^3}{6} - \frac{x^2}{3}(P + wl) + \frac{l^2}{2}\left(\frac{P}{2} + \frac{wl}{3}\right)\right] \tag{5-29a}$$

if $x < \frac{1}{2}l$.

We can determine the deflection for $x > \frac{1}{2}l$ by a similar procedure, using Eqn. (5-23b), except that in this case we eliminate constants of integration using the conditions that at $x = \frac{1}{2}l$, $dy/dx = 0$, and at $x = l$, $y = 0$ (since the beam is supported at this point). Then

$$y = \frac{1}{4EI}\left[\frac{wx^4}{6} + \frac{x^3}{3}(P - wl) - Plx^2 + \frac{l^2x}{2}\left(\frac{wl}{3} + \frac{3P}{2}\right) - \frac{Pl^3}{12}\right] \tag{5-29b}$$

if $x > \frac{1}{2}l$.

As would be expected, both of these expressions lead to the same value for y at the centre of the beam. Substituting $x = \frac{1}{2}l$ in either gives y_0, the central deflection, as

$$y_0 = \frac{l^3}{96EI}\left(2P + \frac{5wl}{4}\right) \tag{5-29c}$$

The quantity I can be determined from the geometry of the cross section of the beam, using Eqn. (5-21). Suppose, for example, that the cross section

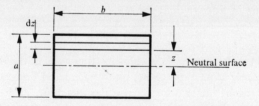

Fig. 5-20 Derivation of second moment of area for rectangular cross section

was rectangular, of thickness a and breadth b, as shown in Fig. 5-20. Then, from Eqn. (5-21),

$$I = b\int_{-a/2}^{a/2} z^2\, dz$$

which reduces to $I = ba^3/12$.

Note that this example would have been simplified by taking $x = 0$ at the point where the force P is applied. The system would then have been symmetrical about the y axis and it would only have been necessary to determine the deflection of one half of the beam. The method used was chosen to emphasize, in a fairly simple problem, the important point that it is generally necessary to obtain a separate solution for each portion of the beam between points where the loading changes discontinuously. The separate solutions are then linked together using the fact that, at the discontinuities in loading, the slope and displacement of the beam are continuous.

Consider now the example shown in Fig. 5-18, in which one of the reaction forces is unknown. We will need to know the slope and deflection of the beam at certain points in order to eliminate constants of integration. Since we know that at the clamp $(x = 0)$ both the slope and deflection of the beam are zero, we will first find the deflection of the portion to the left of A.

From Eqns. (5-25c) and (5-28)

$$EI \frac{d^2 y}{dx^2} = \frac{P(x^2 - 22xl + 21l^2)}{20l} + \frac{R_2(3x - 2l)}{3}$$

We integrate this equation twice (from the boundary conditions given above the constants of integration are zero) giving

$$2EIy = \frac{P(x^4 - 44x^3 l + 126l^2 x^2)}{20l} + \frac{R_2(x^3 - 2lx^2)}{3} \qquad (5\text{-}30a)$$

We can determine R_2 using the fact that the beam is supported at $x = l_1$, and hence at this point $y = 0$. Making this substitution and solving for R_2, remembering $l_1 = \frac{2}{3}l$, gives

$$R_2 = 437P/240 \qquad (5\text{-}30b)$$

Substituting in Eqn. (5-30a) for R_2:

$$1440EIyl = Px^2(6x^2 + 173xl - 118l^2) \qquad (5\text{-}31a)$$

We find the deflection of the right-hand portion of the beam by substituting the value of M from Eqn. (5-28) into (5-25a) and integrating twice. To find the constants of integration we use the fact that, at $x = l_1$, the slope and deflection are the same as for the left-hand portion. Thus, at $x = l_1$ the deflection is zero and we find the slope by differentiating Eqn. (5-31a). So we derive the equation relating deflection to position along the beam for the right-hand portion:

$$19,440EIyl = P(3x - 2l)(27x^3 - 1,170x^2 l + 3,022l^2 x - 874l^3) \quad (5\text{-}31b)$$

At the end of the beam $x = l$, and hence the depression of the end, y_0, is given by

$$y_0 = 121Pl^3/3,888EI \qquad (5\text{-}31c)$$

5-6-3 Deformation of loaded beam due to shear

We saw in Section 5-6-1 that the method of loading produces a shear force as well as a bending moment. In order to determine the beam deformation due to this shear force, we need to know its distribution over the plane on which it acts. However, in this section we show that, provided certain conditions are met, the deformation due to shear is negligible compared with that due to bending. For this, it is sufficient to assume that the shear force is distributed uniformly over the cross section. If the deformation was not negligible, more precise calculations would be necessary.

Figure 5-21 shows a small element of length dx cut from the beam. If A is its area of cross section, the shear stress is F/A. The angle of shear is γ,

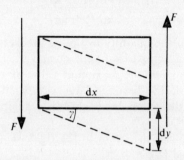

Fig. 5-21 Deformation of element of beam due to shear

and so if dy is the depression of one end of the element relative to the other due to shear, then

$$\gamma = \frac{dy}{dx}$$

and

$$A\mu \frac{dy}{dx} = -F \qquad (5\text{-}32)$$

(from the sign convention defined in Section 5-6-1).

Considering, first of all, the portion of the beam to the right of P in Fig. 5-16, from Eqns. (5-23a) and (5-32),

$$-A\mu \frac{dy}{dx} = \frac{P}{2} + w\left(x - \frac{l}{2}\right)$$

Integrating this equation and eliminating the constant of integration from the boundary condition that at $x = l$, $y = 0$, gives

$$A\mu y_s = \left(P + \frac{wx}{2}\right)(l - x)$$

where y_s is the deflection due to shear. Thus y_{0s}, the deflection due to shear at the centre of the beam, is given by

$$A\mu y_{0s} = \frac{l}{2}\left(P + \frac{wl}{4}\right)$$

If we compare this with the deflection due to bending, given in Eqn. (5-29c),

$$\frac{y_{0s}}{y_0} = \frac{48EI}{\mu A l^2} \frac{(4P + wl)}{(8P + 5wl)}$$

and if the cross section of the beam is rectangular, as illustrated in Fig. 5-20, then

$$\frac{y_{0s}}{y_0} = \left[\frac{4E(4P + wl)}{\mu(8P + 5wl)}\right]\frac{a^2}{l^2}$$

Now the magnitude of the quantity in square brackets is a maximum when wl can be neglected in comparison with P, and since, from Eqn. (5-2), E cannot be greater than 3μ, the maximum value of y_{0s}/y_0 is $6a^2/l^2$. Thus, provided the length of the beam is large compared with its depth, i.e., a^2/l^2 is very small, the deflection due to shear is negligible compared with that due to bending.

Consider the portion of the beam to the right of A in Fig. 5-18. Then, from Eqns. (5-25b) and (5-32),

$$-A\mu\frac{dy}{dx} = \frac{P}{10}\left(\frac{x}{l} - 11\right)$$

If we integrate this equation and eliminate the constant of integration from the boundary condition that at $x = l_1$, $y = 0$, then the depression of the beam due to shear at $x = l$ is

$$y_{0s} = \frac{61Pl}{180A\mu}$$

If we compare this equation with Eqn. (5-31c), which gives the deflection due to bending, we can again show that the shear deflection is negligible compared with that due to bending if a^2/l^2 is very small.

5-6-4 Maximum stress in a loaded beam

We have now seen how to determine the deflection at any point along a beam subjected to a given loading. Another quantity which we sometimes need to know is the maximum stress in the beam. We find this by eliminating R between Eqns. (5-19) and (5-22), giving the expression

$$\sigma_{xx} = zM/I \qquad (5\text{-}33)$$

I is uniform along the beam and so the maximum extensional stress on an element occurs at the position where M is a maximum, and in the element at the greatest possible distance from the neutral axis.

For the portion of the beam to the right of P in Fig. 5-16, the bending moment is given by Eqn. (5-23b). To find the position at which its value is a maximum, we differentiate this equation with respect to x, and equate to zero. Thus, the maximum bending moment occurs at

$$x = \frac{l}{2} - \frac{P}{2w}$$

The value of x given by this equation is less than $\frac{1}{2}$. However, the equation from which it was derived is valid only for values of x between $\frac{1}{2}l$ and l (i.e., it describes the right-hand portion of the beam only), and so the maximum bending moment must occur at $x = \frac{1}{2}l$. Substituting for x in Eqn. (5-23b) gives

$$M_1 = \frac{l}{4}\left(P + \frac{wl}{2}\right)$$

Thus the maximum extensional stress σ_1 occurs in the upper or lower surface (since z is a maximum at these surfaces) at the middle of the beam, and is obtained by substituting the value of M_1 given above in Eqn. (5-33). This gives

$$\sigma_1 = \frac{\pm l z_1(P + \frac{1}{2}wl)}{4I}$$

(z_1 is the value of z at the upper or lower surface of the beam). If the cross section of the beam is rectangular, as in Fig. 5-20, then the maximum extensional stress is given by

$$\sigma_1 = \frac{\pm 3l}{2ba^2}\left(P + \frac{wl}{2}\right)$$

Consider now the problem shown in Fig. 5-18. Combining Eqns. (5-25) and (5-30), we get

$$M = \frac{P(x^2 - 22lx + 21l^2)}{20l} \tag{5-34a}$$

for the right-hand portion of the beam, and

$$M = \frac{P(36x^2 + 519xl - 118l^2)}{720l} \tag{5-34b}$$

for the left-hand portion.

Taking first the portion of the beam to the right of A, if we differentiate Eqn. (5-34a) we find a turning point in the curve of M against x at $x = 11l$. However, the equation is valid only for values of x between $l_1 (= \frac{2}{3}l)$ and l, and hence there is no turning point in this range. The maximum value of bending moment in this portion of the beam is therefore either at $x = \frac{2}{3}l$, or at $x = l$. Substituting in Eqn. (5-34a), we find the maximum is at $x = \frac{2}{3}l$ and is given by

$$M_1 = 61Pl/180$$

Since we want to find the maximum stress in the entire beam, we need to know if the bending moment at any point in the left-hand portion exceeds the above value. By a similar method to that described above, but using Eqn. (5-34b), it can be shown that there is no turning point in the curve of M against x in the range of x in which the equation is applicable ($x = 0$ to $x = \frac{2}{3}l$) and so the maximum value of M must occur at either $x = 0$, or $x = \frac{2}{3}l$. The value at $x = 0$ is given by

$$M_1 = -59Pl/360$$

which is numerically smaller than the value at $x = \frac{2}{3}l$ given above.

Thus, the maximum extensional stress occurs in the element of material in the outer surface at the support A in Fig. 5-18. If the beam is of rectangular cross section, as in Fig. 5-20, this stress is given by

$$\sigma_1 = \pm 61Pl/30ba^2$$

From the above calculations we have determined the maximum extensional stress due to the bending moment. There is also a shear stress, caused by the shear force, but before we can determine the maximum value of this, we need to know how it is distributed over the cross section. It can be shown that the maximum value occurs at the neutral axis and that the value at the upper and lower surfaces is zero (i.e., the opposite of the values of the extensional stress, which are zero at the neutral axis and maximum at the surfaces). However, provided the beam is long and thin, the maximum value of the shear stress is always much less than the maximum value of the extensional stress, and so its calculation will not be considered here.

5-6-5 Elastic energy stored in bending

Figure 5-22 shows a small section of a bent beam. The axis of this section was initially straight and its end faces are perpendicular to this axis. On bending, the faces are inclined at an angle $d\phi$. Then, if ds is the length of the neutral axis and R its radius of curvature,

$$\frac{1}{R} = \frac{d\phi}{ds}$$

Substituting this in Eqn. (5-22) gives

$$d\phi = \frac{M}{EI} ds \tag{5-35}$$

Now suppose that the bending moment is increased to $M + dM$ and the angle between the end faces increased to $d\phi + d\theta$. Then,

$$d\phi + d\theta = \frac{M + dM}{EI} ds$$

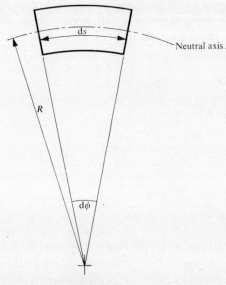

Fig. 5-22

Subtracting Eqn. (5-35) from this gives

$$d\theta = \frac{dM}{EI}\,ds \tag{5-36}$$

This equation relates the increase in the relative rotation of the end faces of the section to the increase in the couple applied to them. Thus, the work done by the couple during this rotation is $M\,d\theta$, and this is equal to the increase in the stored elastic energy resulting from the increase in bending. Hence, if dU is the elastic energy stored in the element ds when the applied bending moment is M, then

$$dU = \int_0^M M\,d\theta$$

Substituting from Eqn. (5-36) for $d\theta$ gives

$$dU = \frac{ds}{EI}\int_0^M M\,dM$$

Hence

$$dU = \frac{M^2\,ds}{2EI}$$

The total elastic energy stored in a beam of length l is therefore

$$U = \int_0^l \frac{M^2\,ds}{2EI} \tag{5-37}$$

If this is applied to the problem shown in Fig. 5-16, EI is uniform along the bar and, since bending is small, $ds = dx$. Thus, the total elastic energy stored in both portions of the beam is given by

$$U = \frac{1}{2EI} \int_0^{l/2} \frac{x^2}{4} [P + w(l - x)]^2 \, dx$$

After integration this gives

$$U = \frac{l^3}{32EI} \left[(P + wl)^2 - \frac{wl}{4}(P + wl) + \frac{w^2 l^2}{20} \right]$$

5-7 The helical spring

5-7-1 Description

A helical spring consists of a wire wound so that a line drawn within it along its axis lies in a helix on a cylindrical surface (Fig. 5-23). The problem is to determine the change in the dimensions of the helix, particularly the length of its axis, produced by a force applied along this axis, and also to determine how this change depends upon the geometry of the helix and the mechanical properties of the wire.

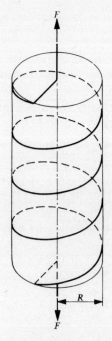

Fig. 5-23 Helical spring

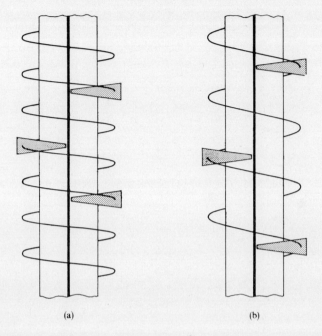

(a)   (b)

Fig. 5-24   (a) Helical spring before loading, (b) Helical spring after loading

5-7-2   Experimental study

The state of stress in an element of the spring is more complicated than the states of stress we have considered so far. So we will first determine by experiment the kind of deformation produced and derive an approximate solution. We will then go on to make a more detailed analysis.

In Fig. 5-24(a) a section of a helical spring is shown before loading. The vertical line marks the axis of the spring, and pointers to this line are fixed to the wire of the spring. The pointers are cemented to the wire, and so they will rotate if the cross section to which they are attached rotates. Figure 5-24(b) shows the same portion of the spring after loading. Note that the pointers still only just touch the line marking the axis of the spring, so the radius of the coils has altered very little, if at all. Also, the pointers are still horizontal, showing that the cross section of the wire has not rotated.

We can now study the deformation of the material of the wire by constructing a model which reproduces these features of the deformation of the spring [Fig. 5-25(a)]. The surface of a length of rubber hose is marked with lines parallel to its axis and with circumferential lines, forming a rectangular grid. The hose is wound into a helix, and fixed by horizontal retort clamps to a support rod lying along the axis of the helix. The coil is extended by

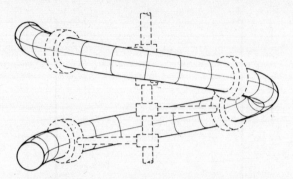

Fig. 5-25   (a) Model of helical spring, unextended

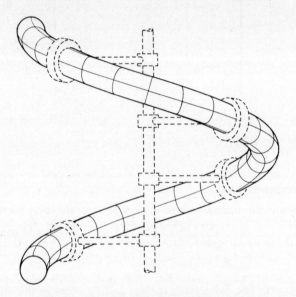

Fig. 5-25   (b) Model of helical spring, extended

increasing the separation of these clamps so that they are kept horizontal and so that their horizontal displacement from the support rod is maintained [Fig. 5-25(b)]. The deformation of the grid on the surface of the hose then corresponds to the deformation of the wire of a helical spring.

A small portion of the grid is shown enlarged in Figs. 5-26(a) and 5-26(b). A piece of right-angled card, held at the intersection of the axial and circumferential lines, shows that the angle between these lines changes on extension of the helix. So a small element of the wire of the spring will be deformed in the way shown in Fig. 5-27. We can see that this deformation is the same as that which occurs during the torsion of a straight wire.

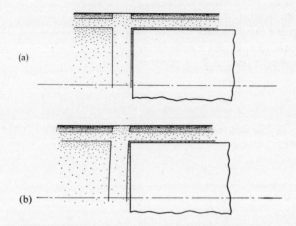

Fig. 5-26 Change in angle between lines of rectangular grid on extension of helix

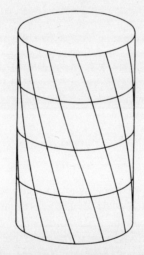

Longitudinal lines were parallel to the axis
before deformation

Fig. 5-27 Deformation of grid marked on surface of wire

5-7-3 Approximate solution

Now we have seen how an element deforms, we could use the method established in the preceding sections. However, in this case, forces on different elements are not parallel to each other and the angles between them are awkward to determine, so solution by the previous method will not be easy. We can avoid the difficulty by considering the stored elastic energy.

From Eqn. (5-13b), the elastic energy stored in a wire of length l and radius a, subjected to a torque T, is given by

$$U = \frac{lT^2}{\pi\mu a^4}$$

From Fig. 5-28, the torque T is produced by the axial force F acting at a

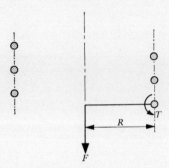

Fig. 5-28   Torque on cross section of wire of helix

distance R from the wire, where R is the radius of the helix, i.e.,

$$T = FR \tag{5-38}$$

Substituting from Eqn. (5-38) in (5-13b)

$$U = \frac{lF^2R^2}{\pi\mu a^4} \tag{5-39}$$

Now let L be the length of the axis of the helix when the axial force is F. Suppose F is increased to $F + dF$ causing L to increase to $L + dL$. The work done by the force will be $F\,dL$ and, since decreasing the force will restore the original conditions, this work must be stored as elastic energy.

The increase in stored elastic energy is $(dU/dF)\,dF$ and since R has been shown experimentally to be constant during deformation, differentiating Eqn. (5-39) we derive the equation

$$\frac{dU}{dF} = \frac{2lFR^2}{\pi\mu a^4}$$

Therefore, equating $F\,dL$ to $(dU/dF)\,dF$ gives

$$\frac{dL}{dF} = \frac{2lR^2}{\pi\mu a^4} \tag{5-40}$$

The term dF/dL is the slope of the load–extension curve of the spring, and is known as the stiffness. Since the right-hand side of Eqn. (5-40) contains no quantity depending on F or L, this slope is constant during extension. Thus, if an axial force F is applied, the length of the axis of the helix increases by an amount L', given by

$$L' = \frac{2lR^2}{\pi\mu a^4}F$$

i.e., L' and F are proportional to each other.

If the helix is close-coiled, i.e., if the axis of the wire is very nearly normal to the axis of the helix, the length of the wire is given by

$$l = 2\pi nR$$

where n is the number of turns on the helix. Thus,

$$L' = \frac{4nR^3F}{\mu a^4} \tag{5-41}$$

5-7-4 Exact solution

We based the approximate solution on the experimental study of Section 5-7-2, and although it shows the most important features it may have neglected others. We now go on to work out an exact solution in which we determine the state of stress in an element of wire. We do this, as previously, by finding the internal forces which must act on the faces of the element to keep it in equilibrium with the external forces.

The element of wire is shown in Fig. 5-29. The cylindrical surface in this diagram, marked with broken lines, is the surface containing the axis of the wire. The element is formed by cutting through the wire along a plane normal to its axis. Since this axis is not normal to the axis of the cylinder, the plane will not contain the line AB (which is drawn on the surface of the cylinder, parallel to its axis). A line

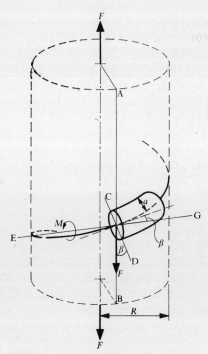

Fig. 5-29 Internal forces on cross section of helix wire

CD, passing through the points at which the circumference of the plane surface cuts the cylindrical surface, is inclined to AB at an angle β, equal to the helix angle. (The helix angle is the angle between the axis of the wire at a given point and the tangent EG to the cylinder at this point, the tangent lying in a plane normal to the axis of the cylinder.)

Consider the internal forces exerted on the face of the element by the material lying to its left. These forces must comprise a force F, acting along AB, to maintain translational equilibrium with the externally applied force, and a couple M (equal to FR) to maintain rotational equilibrium. This couple acts about EG, which is at right angles to the line AB and tangential to the cylindrical surface.

The situation is shown more clearly in Fig. 5-30 which shows a section of Fig. 5-29 on a plane surface tangential to the cylindrical surface along the

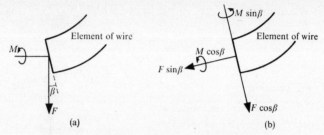

Fig. 5-30   Resolved components of internal forces

line AB. The couple M and force F are shown in Fig. 5-30(a), and Fig. 5-30(b) shows their components along, and perpendicular to, the axis of the wire.

The couple $M \cos \beta$ will cause torsion in the wire and this is the couple considered in the previous approximation. The couple $M \sin \beta$ will cause bending of the wire and will change the radius of the helix. The force $F \sin \beta$ will extend the wire in simple elongation and the force $F \cos \beta$ will shear it. The resultant deformation of the helix is the resultant of the deformations caused by each of these components. We assume that they are the only internal forces acting. (There will, of course, be equal and opposite forces acting on the opposite face.)

(a) Deformation due to torsion and bending   From Eqn. (5-39), since the couple producing torsion has a torque $FR \cos \beta$, the stored elastic energy due to torsion is given by

$$U_{\mathrm{T}} = \frac{l F^2 R^2 \cos^2 \beta}{\pi \mu a^4} \tag{5-42}$$

From Eqn. (5-37) we can calculate the elastic energy stored in a beam of length l subjected to a bending moment M. In the present application, the bending moment M is uniform along the wire and is equal to $FR \sin \beta$. Thus the stored elastic energy due to bending is given by

$$U_B = \frac{lF^2R^2 \sin^2 \beta}{2EI}$$

For a wire of circular cross section, radius a, the second moment of area I about a diameter is $\frac{1}{4}\pi a^4$, and hence

$$U_B = \frac{2lF^2R^2 \sin^2 \beta}{\pi E a^4} \tag{5-43}$$

The total stored elastic energy U is given by

$$U = U_T + U_B$$

and so, from Eqns. (5-42) and (5-43),

$$U = \frac{lF^2R^2}{\pi a^4}\left(\frac{\cos^2 \beta}{\mu} + \frac{2 \sin^2 \beta}{E}\right) \tag{5-44}$$

Now, as in Section 5-7-3, if an increase in force dF causes an increase in the axial length of the helix dL, then the work done by the force is $F\,dL$, which is equal to the increase in the stored elastic energy $(dU/dF)\,dF$.

We can find dU/dF by differentiating Eqn. (5-44), but we now see from above that a couple, $M \sin \beta$, is developed which changes R.

Also the helix angle β will change during extension. Thus, to be completely accurate, we should not regard either R or β as constants when differentiating Eqn. (5-44) with respect to F. However, in this case we can treat them as constants, thereby restricting our work to fairly small extensions of the helix. Differentiating Eqn. (5-44) and equating $F\,dL$ to $(dU/dF)\,dF$ gives

$$\frac{dL}{dF} = \frac{2lR^2}{\pi a^4}\left(\frac{\cos^2 \beta}{\mu} + \frac{2 \sin^2 \beta}{E}\right) \tag{5-45}$$

Hence, regarding R and β as constants, the extension of the spring due to both bending and torsion is still proportional to the applied force. The length of the wire l is given by

$$l = 2\pi n R \sec \beta$$

and, substituting above, gives the extension L' of the helix due to an axial force F as

$$L' = \frac{4nR^3 \sec \beta}{a^4}\left(\frac{\cos^2 \beta}{\mu} + \frac{2 \sin^2 \beta}{E}\right)F \tag{5-46}$$

If β is small, i.e., if the spring is close-coiled, $\cos \beta$ and $\sec \beta$ are nearly unity, and $\sin \beta$ is nearly zero. In this case, Eqn. (5-46) reduces to (5-41), the approximate solution derived from the experimental analysis of the deformation.

(b) Deformation due to simple elongation and shear In addition to the torsion and bending couples, an extensional force of $F \sin \beta$ and a shearing force of $F \cos \beta$ act upon the wire. These cause the dimensions of the helix to change and we should have taken them into account in the above calculation. But we will show that, provided certain conditions are satisfied, their contribution to the deformation is negligible.

To do this we should determine the distribution, over the cross section of the wire, of the extensional and shear stresses due to these forces. A rough calculation will, however, suffice to show their effects to be negligible, and for this we can assume the stresses to be uniform.

The stored elastic energy per unit volume of a wire in simple elongation due to a stress σ_1 was given in Eqn. (5-5c). The stored elastic energy u_2 due to a simple shear stress σ_2 can be derived similarly, and is given by

$$u_2 = \frac{\sigma_2^2}{2\mu} \tag{5-47}$$

Thus, the total stored elastic energy U due to both these stresses is given by the sum of Eqns. (5-5c) and (5-47):

$$U = u_1 + u_2 = \frac{l\pi a^2}{2}\left(\frac{\sigma_1^2}{E} + \frac{\sigma_2^2}{\mu}\right) \tag{5-48}$$

Assuming the stresses to be uniformly distributed

$$\sigma_1 = \frac{F \sin \beta}{\pi a^2} \qquad \sigma_2 = \frac{F \cos \beta}{\pi a^2}$$

Substitution in Eqn. (5-48) gives

$$U = \frac{l}{2\pi a^2}\left(\frac{\sin^2 \beta}{E} + \frac{\cos^2 \beta}{\mu}\right)F^2$$

Equating the increase in stored elastic energy, due to an increase in the axial force, with the work done by that force gives

$$\frac{dL}{dF} = \frac{l}{\pi a^2}\left(\frac{\sin^2 \beta}{E} + \frac{\cos^2 \beta}{\mu}\right)$$

If we compare this with Eqn. (5-45), which gives the value of dL/dF due to bending and torsion, we see that, very approximately,

$$\left(\frac{\mathrm{d}L}{\mathrm{d}F} \text{ in bending and torsion}\right) = \frac{2R^2}{a^2}\left(\frac{\mathrm{d}L}{\mathrm{d}F} \text{ in shear and extension}\right)$$

Provided that the radius R of the helix is large compared with the radius a of the wire, which is true for most wire springs, $2R^2/a^2$ will be very large. Hence, the value of $\mathrm{d}L/\mathrm{d}F$ due to shear and extension will be very small compared with its value due to bending and torsion, and so the contribution of shear and extension of the wire to the total extension of the helix will be negligible.

5-7-5 Worked example

A spring consists of 10 turns of steel wire of radius 1 mm wound into a helix of radius 2 cm. The length of the helix is 20 cm. Determine (a) the helix angle, (b) the stiffness of the spring, (c) the extension due to torsion at an axial load of 10^5 dyn, (d) the extension due to bending at an axial load of 10^5 dyn, and (e) the total stored elastic energy at an axial load of 10^5 dyn. ($E = 20 \times 10^{11}$ and $\mu = 8 \times 10^{11}$ dyn/cm².)

(a) Let a length of helix $\mathrm{d}s$ make a projection $\mathrm{d}y$ on the helix axis, and a projection $\mathrm{d}x$ along the circumference of a circle whose plane is normal to this axis and whose radius is equal to the helix radius. Then, from Fig. 5-31,

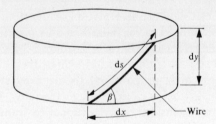

Fig. 5-31   Determination of helix angle

if β is the helix angle

$$\frac{\mathrm{d}y}{\mathrm{d}x} = \tan \beta$$

For one complete turn, $\mathrm{d}x = 2\pi R$ and $\mathrm{d}y = 2$ (since there are 10 turns on an axial length of 20 cm). Hence, since $R = 2$,

$$\tan \beta = 0{\cdot}159 \qquad \text{and} \qquad \beta = 9°2'$$

(b) From Eqn. (5-45), the reciprocal of the stiffness is given by

$$\frac{2lR^2}{\pi a^4}\left(\frac{\cos^2 \beta}{\mu} + \frac{2 \sin^2 \beta}{E}\right)$$

where $l = 2\pi nR \sec \beta$. Substituting given values in this equation, we get

$$\text{stiffness} = 2{\cdot}49 \times 10^5 \text{ dyn/cm}$$

(c) From Eqns. (5-46) and (5-42), the extension due to torsion is given by

$$\frac{2lR^2 F \cos^2 \beta}{\pi \mu a^4}$$

Substituting the necessary values gives

$$\text{extension due to torsion} = 0{\cdot}396 \text{ cm}$$

(d) From Eqns. (5-46) and (5-43), the extension due to bending is given by

$$\frac{4lR^2 F \sin^2 \beta}{\pi E a^4}$$

Hence,   extension due to bending $= 0{\cdot}008$ cm

(e) The stored elastic energy is obtained by substituting the appropriate numerical values in Eqn. (5-44) giving

$$\text{stored elastic energy} = 2{\cdot}01 \times 10^{-3} \text{ J}$$

5-8   Summary

(i) In order to determine the deformation of structures under given loading conditions, we often need to use Saint-Venant's principle (Section 5-2).

(ii) Simple elongation is defined in Section 5-3-1. Under such loading:

(a) stress and strain are uniform throughout the rod (Section 5-3-2),

(b) the axial strain is proportional to the axial stress, the constant of proportionality being called Young's modulus, E (Section 5-3-5),

(c) axial strain (extension) is proportional to transverse strain (contraction), the constant of proportionality being called Poisson's ratio, v (Section 5-3-5),

(d) E and v are related to k and μ by the equations

$$1/E = 1/3\mu + 1/9k$$

$$v = (3k - 2\mu)/(6k + 2\mu) \qquad \text{(Section 5-3-5)}$$

(e) the stored elastic energy, u (Section 5-3-6), per unit volume is given by

$$u = \tfrac{1}{2}E\varepsilon^2 \qquad \text{or} \qquad u = \tfrac{1}{2}\sigma\varepsilon \qquad \text{or} \qquad u = \sigma^2/2E$$

(iii) When a couple T acts in the plane of the cross section of a shaft of circular cross section, the deformation produced is as follows:

(a) the length of the shaft is unaltered (Section 5-4-4),

(b) the radius of the shaft is unaltered (Section 5-4-4),

(c) elements of the shaft are deformed in simple shear (Section 5-4-2),

(d) stress and strain are non-uniform in the cross section of the shaft (Section 5-4-4),

(e) one end of the shaft rotates through an angle ϕ with respect to the other (Section 5-4-4),

(f) the quantity T/ϕ is called the torsional rigidity and is given by $\pi\mu a^2/2l$ for a shaft of length l and radius a (Section 5-4-4).

(iv) (a) The axis of a beam is bent into the arc of a circle by a couple acting in a plane containing the axis. This couple, M, is called the bending moment (Section 5-5-4).

(b) Elements of the beam parallel to the axis are in simple elongation (Section 5-5-2).

(c) A surface exists in which the extensional strain is zero. This is called the neutral surface (Section 5-5-2). If the applied forces have no component along the axis of the beam (pure bending), the neutral surface contains the beam axis which is then called the neutral axis (Section 5-5-4).

(d) The radius R of the neutral axis is given by $M = EI/R$, where I is the second moment of area of the cross section about the line formed by its intersection with the neutral surface (Section 5-5-4).

(e) If y is the displacement of the neutral axis from its undeformed position at a point distance x along the beam axis, then $1/R = \mathrm{d}^2y/\mathrm{d}x^2$ for small bending (Section 5-6-2).

(f) In determining the deflection of a loaded beam, separate solutions must be found for each section of the beam between points where the load changes discontinuously. These can be linked together using the condition that, at such a point, the slope and displacement of the neutral axis are continuous (Section 5-6-2).

(g) Typical methods of loading beams lead to shear forces as well as bending moments, and both vary from point to point along the beam axis. If the depth of the beam is negligible compared with its length, then the deflection due to shear is negligible compared with that due to bending (Section 5-6-3).

(h) The extensional stress (and strain) is non-uniform. It is a maximum at the points in the beam most distant from the neutral surface, and in the cross section where the bending moment is a maximum (Section 5-6-4).

(v) If a helical spring is deformed by a force F acting along the axis of the helix, then for fairly small extensions L' of this axis:

(a) the radius of the helix remains constant (Section 5-7-2),

(b) the wire of the helix is deformed in torsion, bending, simple elongation, and shear (Section 5-7-4),

(c) provided the helix is close-coiled, and the radius R is large compared with that of the wire, a, then most of the extension of its axis is caused by the torsion and is given by $L' = 4nR^3F/\mu a^4$, where n is the number of turns on the helix [Section 5-7-4, and Eqn. (5-41)].

EXERCISES

5-1 A cylindrical bar of density ρ has a length l when it is uniformly supported along its length in a horizontal position. Determine the increase in this length when it is hung vertically from its upper end. If it is pivoted at its centre of gravity in the horizontal position, from where does it gain its stored elastic energy when it is rotated to the vertical position?

5-2 A bar of circular cross section, and of length l, is slightly tapered so that at one end its radius is r, and at the other it is pointed. It is mounted with its axis vertical and the pointed end downwards, and is supported only by its upper end. If the density of the bar is ρ and the Young's modulus is E, determine the axial stress and strain at any point in the bar due to its own weight. (Assume that the deformation of an element is the same as if the cross section were uniform.)

5-3 A structure comprises two cylindrical bars of the same length mounted with their axes parallel and separated by a distance d cm. They are rigidly connected to each other at their ends. One bar is of material with Young's modulus E_1, and is of radius r_1; the other is of material with Young's modulus E_2 and is of radius r_2. The structure is rigidly clamped at one end and a force F, parallel to the axes, is applied at the other end. Determine the point of application of this force with relation to the axes of the bars if they are to extend without bending.

5-4 A motor with power output W erg/s has a cylindrical drive shaft which rotates at n rev/s. If the shaft is of length l cm and is constructed from a material which will withstand a maximum shear strain ε_1, determine the minimum radius which the shaft can have,
  (a) when it is solid,
  (b) when it is a tube with inner radius equal to half the outer.

5-5 A composite shaft has a cylindrical core of radius r_1 cm constructed from a material of shear modulus μ_1 dyn/cm², and an outer tube with inner and outer radii of r_1 cm and r_2 cm respectively, made from a material of shear modulus μ_2 dyn/cm². It is subjected to a torque T dyn-cm. Determine the relative rotation of cross sections at a distance l cm apart, assuming all radial lines in a cross section remain straight and rotate through the same angle. Determine the stored elastic energy in each portion of the shaft.

5-6 A cylindrical shaft of radius r cm passes snugly through a hole of length l cm which is bored through a block. The frictional forces between the shaft and the block are f dyn/cm². Determine the relative rotation of the cross sections at the edges of the block when the shaft is subjected to a torque T dyn-cm. Assume elements deform as in simple twisting.

5-7 A uniform heavy bar AB is supported on a knife edge at C, where AC = AB/4. It is kept in equilibrium by a second knife edge placed over the bar at A at such a height that the free end B is just horizontal. Find the difference in height between the knife edges at A and C.

5-8 A weightless beam is clamped horizontally at one end and a force W is applied to the free end. Show that the depression at a point P, at a distance x from the free end, is the same as the depression at the free end when the same force is applied at the point P. Compare the stored elastic energy in each case.

5-9 A weight W dyn is dropped from a height h cm onto the free end of a bar of length l cm and square cross section, which is clamped in a horizontal position at the other end. If the material can withstand a maximum extensional strain of ε_1, determine the minimum possible thickness of the bar, which may be assumed weightless.

5-10 Young's modulus for a rod can be determined by experiments either in simple elongation or bending. If the values obtained from the two experiments do not agree, what can you conclude about the structure of the rod?

5-11 A wire of radius 0·1 cm is wound into a helix of radius 1 cm. The helix length is 10 cm, and 20 turns are wound onto it. An axial force of 10^6 dyn is applied. If the axial extension of the helix is 1 cm, determine the shear modulus of the wire, given that when the wire is straight the same axial force produces an extension of 2×10^{-3} cm. Find also the maximum angle of shear in the wire.

If the material can withstand a maximum angle of shear of 5×10^{-3}, determine approximately the maximum axial force which can be applied, neglecting the helix angle and its variation with extension. Estimate roughly the effect of this approximation on the value obtained for the force.

6

Experimental validity of basic assumptions

6-1 Introduction

In the preceding chapters we related the deformation of solid bodies to the forces producing it, and to do this we had to make two assumptions about the behaviour of materials under stress. First, we assumed that strain is proportional to the stress, and second, that the strain is completely determined by the stress. In particular, the second assumption implies that the time scale of the experiment does not affect the stress–strain relationship. Also, we restricted our work to small strain and elastically isotropic materials.

In this chapter, we test the validity of these assumptions by experiments on real materials. We restrict the experiments to elastically isotropic materials, at small strain, so that the experimental conditions conform to the restrictions of the theory. As well as studying the validity of the assumptions, we also need to ask whether or not these restrictions are realistic. So we have to consider the results of additional experiments to show the maximum strains to which materials may be submitted, and to indicate whether or not materials commonly occurring are isotropic.

We consider five different types of material, taking one example from each of the following classes:

Crystalline matter   In material of this type, the atoms are packed in a regular pattern. Metals belong to this class and are used in our experiments.

Inorganic glass   This material is unlike crystalline matter in that the atoms are not arranged in a precisely regular pattern. Window and optical glass are common examples.

Polymeric matter   The distinguishing feature of this type of material is that, unlike the other two classes, the molecules are very large. They can be likened to long chains, each link in the chain corresponding to an identical chemical group. The molecules thus comprise a large number of units (often about 5,000) joined end to end, and molecular weights greater than 100,000 are common. We can get some idea of the shape of the molecule from the fact that if its width was to be increased to that of a string 1/20 in. thick, its length would become about 15 yards. For our purpose we can further subdivide polymeric matter into the following three classes. We discuss the difference between these classes at a molecular level in Chapter 8.

1. Rubber

Materials having widely varying properties are all known commercially as rubber, the variation being achieved by the addition of various ingredients. The type of rubber we consider is pure, apart from the addition of chemicals necessary to 'cross-link' the molecules. Cross-linking is the tying together by chemical means of adjacent molecules at a few points along their length, and is essential to avoid permanent dimensional changes during deformation.

2. Polymeric glasses

Whilst materials in this class have the same long chain molecular structure as a rubber, their mechanical properties more closely resemble those of inorganic glasses. Polymethyl methacrylate (Perspex) is a common example.

3. Semicrystalline polymers

This class includes many of the materials commonly known as 'plastic', such as polyethylene, polypropylene, nylon, and polyethylene terephthalate (Terylene).

It is important to realize that the mechanical properties of materials depend greatly upon their exact manufacturing history, and so experimental data given in this chapter should be interpreted with caution. The data will not be quantitatively exact for all samples bearing the same material name, but they will be typical, and our work will reveal various points common to all materials of the same class.

6-2 Proportionality of stress and strain

From Section 5-3, we would expect that, if Hooke's law is obeyed, both axial and transverse strain would be proportional to axial stress when specimens of these materials are stretched in simple elongation. The results of such an experiment are shown in Figs. 6-1. The samples of material were elastically isotropic, and the strains were very small, so that the experimental conditions conformed with the requirements of the theory. For each material straight line graphs are produced.

Note that the axial stress required to produce a given axial strain is very different for different classes of material, being about 5,000 times greater for steel than for rubber. Note also that the curve for glass terminates at a lower strain than the curves for the other materials. This is because glass rods in simple elongation fracture at axial strains little greater than 0·001.

Although the curves for polyethylene and rubber have been drawn as straight lines, there is evidence that this is not strictly correct, and that

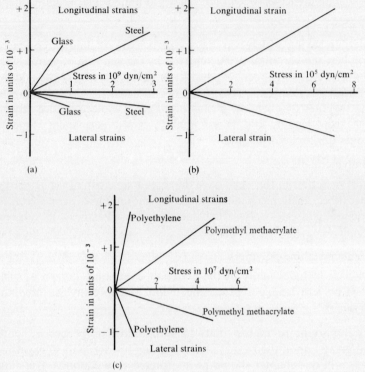

Fig. 6-1 Relationship between axial stress and longitudinal and lateral strains for materials in simple elongation: (a) Glass and steel, (b) Rubber, (c) Polyethylene and polymethyl methacrylate

these materials deviate from linearity. However, *within the range of strains to which small-strain elasticity theory is applicable* this deviation is small—too small to be detectable on the scale to which the graphs have been drawn—and for most purposes can be neglected. Deviation from linear behaviour will also be observed with metals at strains little greater than those at which the graphs in Fig. 6-1 have been terminated.

Hence, on the basis of these experiments, we conclude that, for the classes of material studied, Hooke's law is a valid assumption, provided strains are small enough for elasticity theory to be applicable. It is, however, invalid at strains which are only slightly larger.

6-3   Assumption of perfect elasticity

The assumption of perfect elasticity implies that the stress–strain relationship is unaffected by the time scale of the experiment, or in other words, by the rate at which stress, or strain, is applied. The elastic moduli should therefore be constants, independent of the rate of deformation used in the experiments in which they are measured.

In the next experiments we test this assumption by applying a stress suddenly and then holding it constant. The strain is measured as a function of time, and the ratio of stress to strain is calculated after different lengths of time have elapsed from the moment of application of stress. In a perfectly elastic material this ratio should be independent of the elapsed time. In some of the experiments, the material is deformed in simple elongation, and the ratio is the Young's modulus; in others, the material is deformed in torsion, when the ratio is the shear modulus. However, the results have a similar form for either of these moduli.

The modulus measured after a given period of time is expressed as a proportion of the modulus after a very short time, and this quantity is plotted against elapsed time in Fig. 6-2. The time is plotted on a logarithmic scale, so that the results of experiments completed in less than an hour can be compared with those taking several weeks. Results are given for glass, rubber, polymethyl methacrylate, and polyethylene.

We notice that for each of the three polymers there is a considerable fall in the modulus over a period of an hour (that is, the strain increases, or creeps, with time). Thus the assumption of perfect elasticity is clearly invalid for these materials. For glass, the modulus changes only by about 1 % in a period of 2–3 months. Results are not given for a metal, but, except for very soft metals, such as lead, these would be similar to glass. Thus, while glass and hard metals are not perfectly elastic in the strict sense of the term (since there is a detectable change in modulus with time) they can be assumed to be so for most practical purposes.

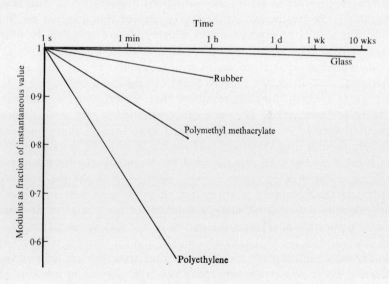

Fig. 6-2   Time-dependence of elastic properties of materials

For glass and metals, the above conclusion is valid only at room temperature. At elevated temperatures the rate of creep is accelerated and can approach that shown for polymers. Also, with metals, the rate of creep will depend upon the magnitude of the strain. If this approaches the limits of the small strain theory, appreciable creep can occur.

The deviation from perfectly elastic behaviour is true only for the shear modulus or the shear component of Young's modulus. Time dependence of the bulk modulus is negligible in all materials.

So we can say that materials are perfectly elastic only in dilatation. The magnitude of deformation with a deviatoric component depends on the *history* of stress application (the length of time for which it has been applied, the rate at which it was increased from zero, etc.), as well as upon the magnitude of the stress. However, provided strains are very small, the temperature is close to normal room temperature, and the experiments do not span an extremely wide time range, this effect will be of importance only for polymers and very soft metals.

In the type of experiment described above, the material deforms under a constant stress (flows). This type of behaviour is characteristic of a *viscous liquid* rather than of an *elastic solid* and is thus termed *visco-elastic*. We discuss ways of describing the stress–strain–time behaviour of such materials in Chapter 7.

6-4   Behaviour at large strain

The experiments described in the preceding sections are confined to very small strains, in order to satisfy the restrictions imposed by our theoretical work. But is this restriction realistic? To answer this question we must consider the results of experiments in which rods are stretched in simple elongation until they break. These results will show whether or not it is possible to deform materials to strains greater than those to which the theory is restricted, and, if so, the type of behaviour exhibited at such strains.

In Figs. 6-3, which show the results of these experiments, axial stress is plotted against axial strain. The small strain definition of axial strain (increase in length divided by original length) is used, even though this is not strictly applicable to strains of the magnitude shown in some of the curves. The stress is taken as the axial force divided by the initial cross-sectional area, and, because the area decreases during extension, it is not the true stress according to the definitions given earlier. Also, in some cases, the area of cross section is not uniform along the length of the test piece at certain stages of deformation. Hence, the curves in Fig. 6-3 are not true stress–strain relationships for the materials chosen, and for this reason stress and strain are shown in inverted commas. (In some texts they are referred to as *engineering stress* and *engineering strain*.) Curves plotted in this way are, however, a convenient means of illustrating the general behaviour.

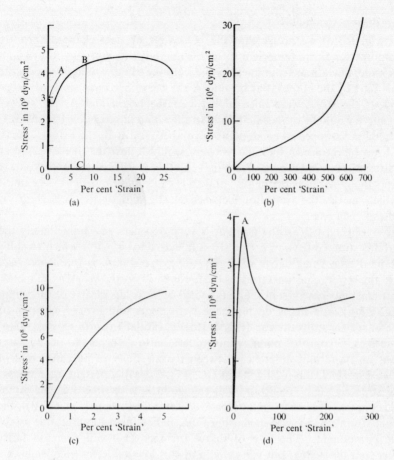

Fig. 6-3 Behaviour of materials in simple elongation at large strain: (a) steel, (b) rubber, (c) polymethyl methacrylate, (d) polyethylene

A curve is not given for glass since this material usually breaks in simple elongation at very small strains. However, it is clear from the figures that it is possible to deform all the other materials to large strains. The limitation of the theory to small strain is therefore extremely restrictive, and methods of describing deformation at large strains are required if the response of materials to deforming stresses is to be measured at all possible strains.

Such methods must include the correct form of the stress–strain relationship at large strains, and it is clear from Fig. 6-3 that this is not linear. Furthermore, different types of material differ greatly in behaviour (for example, rubber can be extended to 600% 'strain' without fracture, whereas steel will break at a 'strain' of between 25% and 30%). Let us now consider some of the characteristics of the different types of large strain behaviour.

6-4-1 Plastic strain

From Fig. 6-3(a) we notice that the behaviour of steel changes abruptly at the point A. From the origin to A, the curve is substantially linear (on a larger scale, deviations from linearity, i.e., Hooke's law, would be noticeable on this part of the curve), but beyond A the curve becomes markedly non-linear. If the stress were released between the origin and A, the stress–strain curve would be retraced, and the strain would fall to zero. If, however, the stretching stopped at some point beyond A, say at B, the original curve would not be retraced. Instead, recovery would follow the line BC, and at zero stress the strain OC would remain. In other words, extension beyond A deforms the material permanently, and so is called *plastic strain*.† The point A, which marks the transition between elastic and plastic behaviour, is called the *yield point*.

For steel [Fig. 6-3(a)] the point A is very well defined. Immediately the material is strained beyond A, the stress falls to a lower value which remains constant over a small range of strain. The stress at A is the *upper yield stress*; the lower, constant, stress is the *lower yield stress*. Whereas all metals exhibit plastic behaviour, in many the yield point is less clearly defined than in Fig. 6-3(a), and many do not have both upper and lower yield points.

The curve for polyethylene [Fig. 6-3(d)] is similar in form to Fig. 6-3(a) and, in fact, beyond the point A this material also deforms plastically. But there is an important difference between the plasticity of polymers and of metals. For steel, the plastic strain OC [Fig. 6-3(a)] is present after release of the stress at the point B, and will change little with time, i.e., the visco-elastic component is very small. However, for polyethylene [Fig. 6-3(d)], the residual strain decays with increasing time, indicating an appreciable visco-elastic component. The rate of decay (or recovery) will decrease fairly quickly to a low value, but will remain at this low value for a considerable period, and the extent of the recovery is difficult to determine. In fact, the rate is so low that the dimensions eventually reach values which remain constant on any reasonable time scale, and recovery can be said to be finished. However, even when this state has been reached, various factors (for example, increase of temperature) can cause further recovery, and if the temperature is increased to the crystal melting point complete recovery will occur. Another method of producing further recovery in some polymers is to immerse them in hot water or other liquids. (This type of recovery is the cause of shrinking in clothing.) Such behaviour arises from the long chain structure of polymer molecules, and does not occur in metals (Chapter 8).

---

† In Chapter 7 it is shown that there is a residual strain in a *visco-elastic* material immediately following stress release, and so we must distinguish between *plasticity* and *visco-elasticity*. *Plastic strain* is that strain which remains *permanently*, and which does not change with time. *Visco-elastic strain* will *decrease with increasing time*, eventually falling to zero. A residual strain can have both plastic and visco-elastic components. Such a strain would decrease with increasing time because of the visco-elastic component, but this decrease (or recovery) would eventually stop, leaving a permanent, plastic, strain.

For polyethylene, the portion of the stress–strain curve after point A corresponds to the formation of a constriction, called a *neck*, in the specimen. This constriction deepens until the minimum in the curve is reached, and further deformation occurs by propagation of the shoulders of the neck along the test piece (Fig. 6-4). We can regard the specimen as existing in two

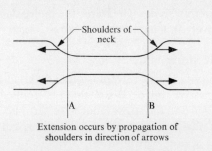

Extension occurs by propagation of
shoulders in direction of arrows

Fig. 6-4 Deformation of a test piece by means of a neck

states—the portion outside the shoulders of the neck, and the portion between the shoulders (AB in Fig. 6-4), which is more highly deformed. Further deformation occurs by a portion of the material in the first state being transformed into the second. Since, in this condition, material in two different states exists in the same test piece, it is obvious that a single 'stress–strain' curve obtained by measuring the load applied to the test piece, and the deformation between its ends [Fig. 6-3(d)], is meaningless as a description of the true relationship between stress and strain for the material. Whether or not a polymer yields in this manner depends upon the temperature and the rate of increase of extension. In the final portion of the stress–strain curve for steel [Fig. 6-3(a)], where the stress is decreasing with increasing strain, a neck is also being formed. However, this does not propagate along the test piece. Instead, it deepens until fracture occurs.

Although there is no pronounced yield point in the curve for polymethyl methacrylate [Fig. 6-3(c)], if stretching were stopped at 3–4% extension, a small amount of plastic deformation would be present. Materials, such as glass and polymethyl methacrylate, which break at strains of a few per cent or less, with little, if any, plastic deformation, are referred to as *brittle* materials. Materials such as steel and polyethylene, which deform plastically by a considerable amount before fracture, are called *ductile*.

Figures 6-3(a)–(d) relate to changes in the length of the test piece, they give no information about volume changes. Had volume changes been measured it would have been found that the permanent dilatational strain (if any) was negligible compared with the deviatoric. Plastic deformation is therefore a *deviatoric* strain, not a dilatational strain.

The study of plasticity is important in many different fields of work. A structural engineer must design structures that will not deform plastically in service (often under more complicated stress systems than simple elongation). On the other hand, shaping operations on both plastic and metal components require permanent deformation, and for these operations methods of calculating the forces necessary to produce large plastic strain are desirable. The physicist is interested in the mechanism of yielding at a molecular level in materials. The first two aspects of this subject are not considered further in this book. The third, the molecular aspect, is discussed briefly and qualitatively in Chapter 8.

6-4-2 Large elastic strain

The behaviour of rubber at large strain is quite different from the behaviour of the materials discussed above. If stretching is stopped at any point along the curve in Fig. 6-3(b) and the stress reduced to zero, no plastic deformation is observed. There might be a small residual strain, but this will disappear with time, and so is a visco-elastic strain. Even if rubber is extended to six or seven times its normal length, the deformation is entirely recoverable. There is no yield point and no region of plastic deformation on its stress–strain curve. Such materials are described as being *highly elastic*, and the deformation is referred to as *large elastic strain*.

Any highly elastic material is usually known as a rubber (or, in American texts, as an elastomer), whatever the chemical constitution of its molecules. Natural rubber was originally the only material having this property, but nowadays a wide range of synthetic, highly elastic, materials are available, with different chemical compositions (but all with long chain molecules).

The large extensional strains which occur in large elastic strain are most conveniently specified by the extension ratio (Section 2-13). For simple elongation, we define the extension ratio as the extended length divided by the unextended length. Thus, an axial strain of 100% corresponds to an extension ratio of 2.

For rubbers in simple elongation, the axial stress (expressed as force per unit cross-sectional area *in the unstrained state*) and the extension ratio α are related by the equation

$$f = G(\alpha - 1/\alpha^2) \tag{6-1}$$

up to extension ratios of about 2 (the symbol f is used because this is not a stress according to the strict definition of the term). This equation is derived from a theory of the molecular mechanism of large elastic strain which is discussed in Chapter 8. The term G has the nature of an elastic modulus. It is independent of the chemical composition of the rubber molecule, but depends instead on the lengths of the molecular chains between cross-links. This is in accord with its theoretical derivation. At larger extension ratios this expression breaks down and the equation

$$f = 2(\alpha - 1/\alpha^2)(C_1 + C_2/\alpha) \tag{6-2}$$

describes the behaviour better. This equation is derived from a generalization of the first four chapters of this book to include large elastic strain in bodies which are isotropic in the unstressed state, and does not depend on the particular molecular mechanism of extension. The terms C_1 and C_2 are constants defining the elastic properties of the material. It can be shown that C_1 is related to G (defined above), but the significance of C_2 in molecular terms remains a puzzle.

If we were to measure the volume changes during the extension of a rubber, we would find that they were negligible compared with the extensional strain. Hence, dilatational strain is very small, and large elastic strain is entirely deviatoric.

6-5   Restriction to isotropic materials

The equations we have developed to relate stress and strain are not only restricted to small strains; they are also restricted to elastically isotropic material. We can study the isotropy of a material in two ways.

First, if the material is available in the form of a rod, we can measure μ, E, and v, experimentally. Now, from Eqns. (5-2) and (5-4),

$$E = 2\mu(1 + v)$$

and thus, *if the material is isotropic*,

$$2\mu(1 + v)/E = 1$$

We can calculate the quantity $2\mu(1 + v)/E$ from our measurements, and if it is different from unity the material is anisotropic. It is therefore called the *anisotropy factor*.

Second, if the material is available in sheet or block form, we can cut strips in various directions and measure their Young's moduli. If different values are obtained for strips cut from different directions, then the material is anisotropic.

Experiments using these methods show that materials in many of the forms in which they are commonly used are anisotropic. For example, textile fibres are highly anisotropic. So, to a lesser degree, are plastics, whether in the form of film, plate, or rod. Some anisotropy often exists in metal sheets.

The reason for this is that the processes by which the material is shaped often introduce directional properties. Metal is a crystalline material and a single crystal would be highly anisotropic. However, a metal block is made up of these small crystals pointing in different directions and is therefore isotropic. But if the block is formed into a sheet by squeezing it through rollers which reduce its thickness and increase its length, the crystals tend to tilt into preferred directions and so anisotropy is introduced.

Synthetic textile fibres, which are semicrystalline polymers, are made by extending filaments of the material plastically to several times their isotropic length. This operation tends to align the molecular chains in a preferred direction and so these materials are highly anisotropic. Similar processes occur in the manufacture of plastic film, and also make it difficult to avoid introducing some anisotropy during the manufacture of rod and plate. Natural textile fibres, such as cotton and wool, are also anisotropic, the alignment being introduced in this case during the growth of the fibre.

We see, then, that many materials in the form in which they are commonly used exhibit anisotropy to some degree. It is therefore necessary to extend the theory relating stress and strain to include anisotropic materials. This has been done, but the mathematics are beyond the scope of this book. It is found that to describe the elastic properties of a completely anisotropic material, 21 independent elastic moduli are necessary (instead of the two required for an isotropic material). If any symmetry exists in the structure this will reduce the number of moduli. For example, in the case of a textile fibre, cylindrical symmetry exists, and this reduces the number to five. In the case of a single cubic crystal, the atoms lie at the corners of a cube, and this symmetry reduces the number to three.

6-6   Heterogeneous materials

In previous chapters we have implicitly assumed that the materials are *homogeneous*, i.e., their elastic properties are uniform throughout the material. However, this is not true for many materials in common use, and these, in which the elastic properties vary from point to point, are called *heterogeneous materials*. Common examples include paper (which comprises a bonded matrix of fibres directed at random), cloth (which comprises fibres woven in two directions at right angles), cord or yarn (which comprises a twisted bundle of fibres), fibre-reinforced plastics, timber, and plywood.

Nearly all materials can, in fact, be regarded as heterogeneous if the scale on which they are examined is small enough. Thus an isotropic metal sheet consists of microscopic crystals oriented at random. If the elastic properties of this sheet are examined on a microscopic scale they will be found to vary from point to point, depending on the orientation of the particular crystal being studied. The heterogeneity of a material such as this is unimportant in determining the deformation of a large piece. It is sufficient to use the average elastic properties.

If, however, the size of the heterogeneous parts is comparable with the size of the whole piece, this is not possible. In such cases we must regard the body as a structure made up of component parts, and determine the force–deformation relationship of the structure from those of its parts. This is often extremely difficult, and the elastic properties of many structures have yet to be determined in this way.

6-7   Summary

The experimental results discussed in this chapter lead to the following conclusions.

(i) Hooke's law is obeyed by all materials at very small strains, i.e., those within the range of the theoretical development (Section 6-2).

(ii) Few, if any, materials are perfectly elastic. In experiments covering a reasonable time scale, hard metals and glass at very small strains and room temperature can be treated as such. Polymers under these conditions are markedly visco-elastic, and the simple elasticity theory can be used with these materials only if the moduli are measured in experiments covering a similar time scale to that which will be operative in their application (Section 6-3).

(iii) Most materials can be extended to strains which are too large to use small strain elasticity theory (Section 6-4) and the phenomena of high elasticity (Section 6-4-2) and plasticity (Section 6-4-1) occur at these large strains.

(iv) Many materials are anisotropic in the forms in which they are commonly used (Section 6-5).

7

Description of visco-elastic behaviour

7-1 Introduction

In the previous chapter we saw that materials are not perfectly elastic, but their shear modulus depends upon preceding stress or strain history. It is therefore desirable to be able to describe the visco-elastic properties of a material so that its response to any known stress history can be calculated. In this chapter we develop methods of doing this, which are satisfactory at small strains. This type of description is particularly important for polymers.

7-2 Visco-elastic behaviour of materials

Section 6-3 described an experiment to study visco-elastic behaviour, and gave the results for several materials. In this experiment we observed the phenomenon known as *creep*, and we now go on to examine the creep behaviour of a typical visco-elastic material in more detail, and describe two other experiments which are used to study visco-elasticity.

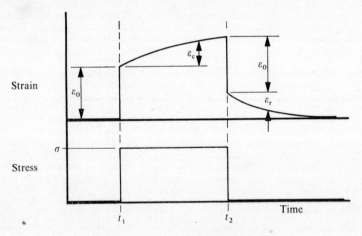

Fig. 7-1 Creep of visco-elastic material

Figure 7-1 shows the creep of such a material. In this figure, stress and strain are plotted against time. (For convenience, we describe the results of a simple elongation experiment, the stress and strain being the axial values. Figure 7-1 and the description which follows would be equally applicable to a torsion experiment.)

A stress σ is instantaneously established at time t_1 and maintained constant until time t_2, when it is instantaneously removed. When the stress is applied, an *instantaneous elastic strain* ε_0 occurs. While the stress is held constant, the strain continues to increase by an amount ε_c, the *creep strain*. When the stress is removed, there is an *instantaneous elastic recovery* ε_0, equal in magnitude to the instantaneous elastic strain. In time, the residual strain will have recovered by an amount ε_r, the *creep recovery*.

The *creep modulus* E_c† is defined as $\sigma/(\varepsilon_0 + \varepsilon_c)$ and the visco-elastic behaviour in creep can be described by expressing E_c as a function of time. This function is often represented graphically. It is sometimes more convenient to express results in terms of the *creep compliance* D_c,‡ instead of the creep modulus. This is simply the reciprocal of the modulus, and is given by $(\varepsilon_0 + \varepsilon_c)/\sigma$. Clearly, the mechanism responsible for creep is also responsible for creep recovery. Hence, any method of visco-elastic analysis should enable recovery behaviour to be calculated from the creep modulus.

In the second type of experiment, we observe the phenomenon of *stress relaxation*. It is conveniently demonstrated by considering simple elongation, but other forms of deformation would serve equally well. In this experiment, the specimen is clamped at a fixed extension, and a stress σ_0 is developed at the instant, t_1, at which the extension is applied. This stress relaxes with time by an amount σ_s (Fig. 7-2). If, at time t_2, the extension is instantaneously

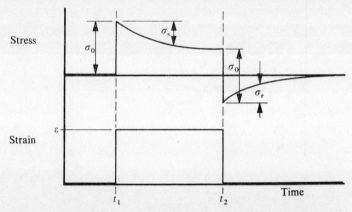

Fig. 7-2   Stress relaxation in visco-elastic material

reduced to zero and the specimen reclamped, the stress falls instantaneously by an amount σ_0, causing a compressive stress to be generated. This will relax with time by an amount σ_r.

† We use the symbol E, for Young's modulus, since we are describing a simple elongation experiment. Were we describing a torsion experiment, the shear modulus μ would have been used. The symbol G is also frequently used to denote shear modulus in visco-elastic experiments.

‡ D is the symbol used for the *tensile* compliance, i.e., the reciprocal of the Young's modulus. J is generally used for the shear compliance, or reciprocal of the shear modulus.

The visco-elastic behaviour in stress relaxation can be described by expressing the *stress-relaxation modulus* E_s, defined as $(\sigma_0 - \sigma_s)/\varepsilon$, as a function of time. However, it is sometimes more convenient to use the *stress-relaxation compliance* D_s, which is simply the reciprocal of the modulus. The recovery from stress relaxation is caused by the same process that causes the relaxation, and so it should be possible to calculate this recovery from the variation of of the stress-relaxation modulus with time. Also, creep and stress relaxation must both be caused by the same process, and so, if a material is characterized in one experiment, it should be possible to predict its behaviour in the other.

In the third type of experiment, a strain which varies sinusoidally with time is applied to the material, and the stress measured as a function of time. Graphs of stress and strain against time, similar to those of Fig. 7-3(a), are obtained.

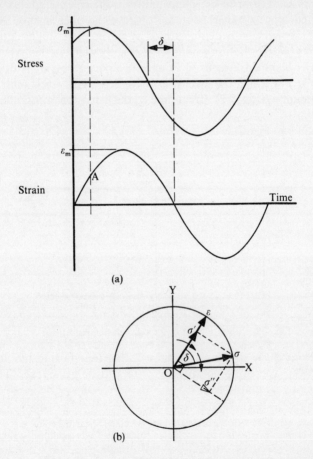

Fig. 7-3 Response of a visco-elastic material to sinusoidally varying strain

We see from Fig. 7-3(a) that the stress varies sinusoidally with time, but is out of phase with the strain. Thus, we need two parameters to specify the response of the material to this type of strain history; one to give the relationship between the amplitudes of the stress and strain curves, the other to give the relationship between their phases. The first is the *absolute modulus* $|E|$ $(= \sigma_m/\varepsilon_m)$ and the second is the phase angle δ. (The phase angle is obtained by dividing the time scale representing one complete cycle into 2π radians, and determining the number of radians between two identical positions, one on the stress curve, the other on the strain curve.) Instead of the phase angle, the *loss factor*, $\tan \delta$, is often specified. Both $|E|$ and $\tan \delta$ vary with the frequency at which the strain oscillates, and a complete specification of the visco-elastic properties of the material will express them as functions of frequency.

Figure 7-3(b) shows another method of expressing the results of this experiment. Since the strain varies sinusoidally with time, its value at any instant can be represented by the projection of the rotating vector ε (of magnitude ε_m) on the axis OX. This vector rotates in a clockwise direction with angular velocity $2\pi n$ radians per second, where n is the frequency of the strain oscillation. In Fig. 7-3(b) it is at a position representing the strain at the point A in Fig. 7-3(a). The stress can similarly be represented on the same figure by a vector σ rotating with the same angular velocity. Its length will be σ_m and its position as shown in Fig. 7-3(b). Thus, the stress is in advance of the strain by the phase angle δ. The ratio of these two vectors (i.e., σ/ε) is the *complex modulus* E^*. (It is called complex because—as will be shown—it can be represented by complex number notation.)

The vector σ can be resolved into two components, σ' along ε (in phase), and σ'' at right angles to ε (out of phase). So we can now define two moduli: the *in-phase modulus* E', given by σ'/ε; and the *out-of-phase modulus* E'', given by σ''/ε. Since

$$\sigma' = \sigma \cos \delta \qquad \text{and} \qquad \sigma'' = \sigma \sin \delta$$

$$E' = \frac{\sigma'}{\varepsilon} = \frac{\sigma \cos \delta}{\varepsilon} = E^* \cos \delta$$

and, similarly,
$$E'' = E^* \sin \delta$$

Thus these moduli are only an alternative way of expressing the experimental results; they do not provide any further information.

Using the normal notation of complex numbers, $\sigma = \sigma' + i\sigma''$, since σ is the vector sum of σ' and σ''. Thus

$$E^* = \sigma/\varepsilon = (\sigma' + i\sigma'')/\varepsilon = E' + iE''$$

Clearly, the moduli determined from this type of experiment are related to the creep and stress-relaxation moduli, and their variation with time.

It should be possible, therefore, to characterize the visco-elastic properties of a material in a way that will enable its response to these stress or strain histories to be predicted.

7-3   Use of models to describe visco-elastic behaviour

Consider the creep of a visco-elastic material. The deformation which occurs immediately the load is applied corresponds to that which would occur in a perfectly elastic material, while the deformation which occurs under a steady load corresponds to that of a viscous liquid. This suggests that it might be possible to reproduce the response of the material to this stress history with a model made up of elastic and viscous components. Further, such a model might respond to any other stress history in an identical way to the material. If so, we would be able to describe the visco-elastic properties of the material by the moduli and viscosities of the components of the model.

7-3-1   Maxwell model

In 1867, Clerk Maxwell suggested using the model shown in Fig. 7-4, and this possibility has been explored by many people. The model comprises a

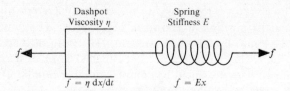

Fig. 7-4   Maxwell model

spring and dashpot† connected in series. The spring has a stiffness, E, such that a force, f, will produce an extension, x, given by

$$f = Ex \qquad (7\text{-}1)$$

The dashpot has a viscous coefficient, η, such that a force, f, will cause a rate of extension of $\mathrm{d}x/\mathrm{d}t$, which is given by

$$f = \eta \, \mathrm{d}x/\mathrm{d}t \qquad (7\text{-}2)$$

† A dashpot is a mechanical device which extends at a constant velocity when subjected to a constant force, and consists of a piston moving in a cylinder filled with a viscous fluid. The viscous coefficient of a dashpot is defined as the ratio of the force to the velocity of movement of the piston (i.e., the rate of extension). The symbol η is used for this, but it should be noted that the same symbol is also frequently used for viscosity, which is dimensionally different. Similarly, the symbol E is used for the stiffness of the spring and is also frequently used for Young's modulus, which again is dimensionally different. Confusion will not arise provided it is remembered that, when referring to models, η and E represent the viscous coefficient and stiffness respectively, but when referring to materials they represent the viscosity and modulus.

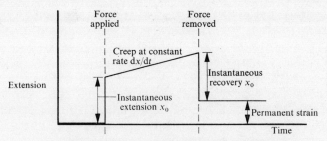

Fig. 7-5 Response of a Maxwell model to creep

Let us now perform a creep experiment on this model. When the force f is applied, the spring extends instantaneously by an amount x_0 (Fig. 7-5). As f is held constant, the dashpot extends at a constant rate dx/dt, the extension of the spring remaining constant. When the force is removed, the spring recovers instantaneously, its extension decreasing by x_0, but the dashpot does not recover, since there is now no force acting on it. Thus, as shown in Fig. 7-5, the model demonstrates all the observed experimental behaviour except creep recovery.

7-3-2 Kelvin or Voigt model

The model shown in Fig. 7-6 was suggested by Kelvin, and also, independently, by Voigt. It comprises a spring and dashpot obeying Eqns. (7-1) and (7-2), but this time they are connected in parallel. Now when the force f is applied the dashpot prevents any instantaneous extension. Under the steady force, creep takes place, the rate being given by

$$f_d = \eta \, dx/dt$$

where

$$f = f_d + f_s \tag{7-3}$$

and

$$f_s = Ex$$

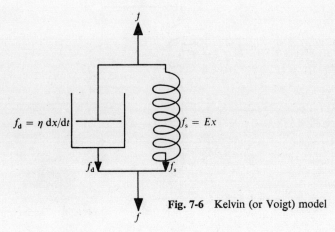

Fig. 7-6 Kelvin (or Voigt) model

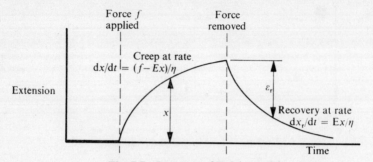

Fig. 7-7   Response of Kelvin model to creep

Thus
$$dx/dt = (f - Ex)/\eta$$
and so, as creep proceeds, since x increases, the rate of creep decreases.

When the force is removed, instantaneous recovery is again prevented by the dashpot. However, Eqn. (7-3) becomes
$$f_d + f_s = 0$$
and, since the spring is extended, f_s is greater than zero; f_d is therefore negative, which means that a force acts on the dashpot to cause recovery. This process is shown in Fig. 7-7, from which we see that the model demonstrates creep and creep recovery, but not instantaneous recovery or extension.

7-3-3   Three-element model

We have seen that the Maxwell model demonstrates all the experimentally observed features of creep except creep recovery; the Kelvin model demon-

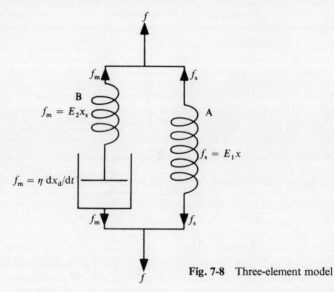

Fig. 7-8   Three-element model

strates all the features except instantaneous extension and recovery. Thus a combination of the two might well reproduce experimental behaviour exactly. This *three-element model* (it is also known as the *standard linear solid*) is shown in Fig. 7-8.

When a force f is applied to this model, the force acting on spring A is f_s, and that acting on the Maxwell element is f_m, where

$$f = f_s + f_m \tag{7-4}$$

Let the total extension of the model be x. This is the extension both of the spring A and the Maxwell element, since they are rigidly connected in parallel. Thus

$$f_s = E_1 x \tag{7-5}$$

The extension of the Maxwell element is divided between spring B (extension x_s) and the dashpot (extension x_d) so that

$$x = x_s + x_d \tag{7-6}$$

Also

$$f_m = E_2 x_s \tag{7-7}$$

and

$$f_m = \eta \frac{dx_d}{dt} \tag{7-8}$$

Consider now the response of this model to a steady force. At the instant the force f is applied, both springs extend instantaneously, but the dashpot remains unextended. Thus, the extension of both springs must be the same, x', and, from Eqns. (7-4), (7-5), and (7-7),

$$x' = \frac{f}{E_1 + E_2}$$

Since there is now a tension in the Maxwell element given by

$$f_m = E_2 x'$$

the dashpot starts to extend at a rate dx_d/dt, causing spring B to contract, and f_m to decrease. This will cause f_s to increase since the sum of f_m and f_s must remain constant at the value f, and so the total extension x of the model increases. In other words, creep occurs.

The rate of creep, dx/dt, can be obtained mathematically. From Eqn. (7-6),

$$\frac{dx}{dt} = \frac{dx_s}{dt} + \frac{dx_d}{dt}$$

and, substituting from Eqns. (7-7) and (7-8),

$$\frac{dx}{dt} = \frac{1}{E_2} \frac{df_m}{dt} + \frac{f_m}{\eta}$$

From Eqns. (7-4) and (7-5) $f_m = f - E_1 x$

and substituting from this in the equation above gives

$$(E_1 + E_2)\frac{dx}{dt} + \frac{E_1 x}{\tau} = \frac{df}{dt} + \frac{f}{\tau} \tag{7-9}$$

To obtain Eqn. (7-9) we make the substitution $\tau = \eta/E_2$, where τ is the *relaxation time* of the Maxwell element. (The reason for this title will become clearer when the behaviour of the model in stress relaxation is considered.) Equation (7-9) is a differential equation from which the force–extension–time relationship of the model can be calculated for any history of force application or extension.

In a creep experiment, $df/dt = 0$. Substituting this condition in Eqn. (7-9) gives

$$\frac{dx}{dt} = \frac{f - E_1 x}{\tau(E_1 + E_2)} \tag{7-10}$$

and this refers only to creep. This equation shows that, since x increases as creep occurs, the rate of creep will decrease with time.

When the force is removed, the springs retract instantaneously by an amount equal to the instantaneous extension. However, since the dashpot has extended during creep, and the total applied force has now been reduced to zero, spring B must be in compression and spring A in extension.

Thus, there is a force tending to make the dashpot contract, reducing the compression in B, which in turn allows A to contract, i.e., creep recovery takes place until the model recovers its original length.

The rate of creep recovery can be found by substituting $f = 0$ in Eqn. (7-10) giving

$$\frac{dx}{dt} = -\frac{E_1 x}{(E_1 + E_2)\tau}$$

Since x decreases as recovery progresses, the rate of creep recovery decreases with time.

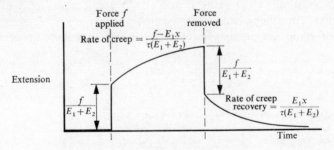

Fig. 7-9 Response of the three-element model to creep

The response of this model is plotted in Fig. 7-9, and if we compare it with Fig. 7-1, we see that the model exhibits, qualitatively, the same creep behaviour as a visco-elastic material.

The response of this model to a stress-relaxation type of experiment will now be considered and shown to be qualitatively the same as that of a visco-elastic material (as illustrated in Fig. 7-2).

If an extension x is instantaneously applied, spring A and spring B both extend by an amount x. Therefore,

$$f_m = E_2 x \quad \text{and} \quad f_s = E_1 x$$

Since f_m and f_s are developed instantaneously, an initial tension f given by

$$f = f_s + f_m = x(E_1 + E_2) \tag{7-11}$$

is developed instantaneously. This corresponds to the stress σ_0 developed at time t_1 in Fig. 7-2.

The extension x applied to the material is now held constant, and so the extension of spring A must remain constant at this value. The tension f_s, therefore, also remains constant. However, the tension in the Maxwell element gradually falls as the tension in spring B extends the dashpot. Therefore, the total tension in the model decreases with time. The rate of decrease can be found from Eqn. (7-9), by substituting $dx/dt = 0$, which is a necessary condition in a stress-relaxation experiment. This gives

$$df/dt = -(f - E_1 x)/\tau \tag{7-12}$$

In this equation df/dt must be negative since, from Eqn. (7-11), f is initially greater than $E_1 x$. Thus, f must relax with increasing time. Furthermore, from Eqn. (7-12), as f decreases the rate of relaxation must decrease. The behaviour of the model thus corresponds to that of the material between times t_1 and t_2 in Fig. 7-2.

When the applied extension is instantaneously reduced to zero, f_s falls to zero, but, since spring B is compressed to compensate for the extension which has occurred in the dashpot, f_m will become negative. This negative tension relaxes as it gradually causes the dashpot to close. The instantaneous fall in tension in the model must be $x(E_1 + E_2)$, since the reduction in extension occurs initially entirely in the springs. Thus, from Eqn. (7-11), the reduction in tension is equal to that instantaneously developed, which again corresponds to the response of the material illustrated in Fig. 7-2. To find the rate of relaxation of the negative tension, we substitute $f = 0$ in Eqn. (7-12), giving

$$df/dt = -f/\tau \tag{7-13}$$

Since f is negative, df/dt is positive, and so the tension approaches zero with increasing time, corresponding to the behaviour shown in Fig. 7-2. As it does so, however, the rate of relaxation decreases. Integrating Eqn. (7-13), we get

$$f = A \exp(-t/\tau)$$

where A is a constant of integration which we can eliminate by supposing that, at some stage during the process represented by this equation, the tension f has the value f_0, and that the time at which this occurs is zero. Then

$$f/f_0 = \exp(-t/\tau)$$

We can use this equation to give a physical definition of the relaxation time, since, when $t = \tau$, i.e., *when a time equal to the relaxation time has elapsed*,

$$f/f_0 = 1/e$$

Thus the relaxation time is *the time required for the tension in a Maxwell element held at constant extension to relax to* $1/e$ *of its original value*.

We have now seen that a three-element model will reproduce qualitatively the behaviour of a visco-elastic material in creep and stress relaxation. We now go on to show that it will also reproduce qualitatively the behaviour of such a material subjected to sinusoidally varying extension.

We can see from a simple physical argument that the tension and extension in the model will be out of phase, for, whereas the tension in the springs is proportional to (and therefore varies in phase with) their extension, the tension in the dashpot is proportional to its rate of extension. For an extension varying sinusoidally with time, the rate of extension (and hence the tension in the dashpot) is 90° out of phase with the extension. The combination of these effects causes the tension and extension to be out of phase, corresponding to the behaviour of the material shown in Fig. 7-3.

We will now express this argument mathematically. Let the extension and time be related by the equation

$$x = x_0 \sin \omega t \tag{7-14}$$

then
$$dx/dt = \omega x_0 \cos \omega t$$

Substituting these in Eqn. (7-9) gives

$$(E_1 + E_2)\omega x_0 \tau \cos \omega t + E_1 x_0 \sin \omega t = \tau df/dt + f$$

This is a differential equation which may be solved to give

$$f = x_0 \left[\frac{E_1^2 + (E_1 + E_2)^2 \omega^2 \tau^2}{1 + \omega^2 \tau^2} \right]^{1/2} \sin(\omega t + \delta) \tag{7-15}$$

where
$$\tan \delta = \frac{E_2 \omega \tau}{E_1 + \omega^2 \tau^2 (E_1 + E_2)}$$

This represents a sinusoidal force–time relationship which leads the extension–time relationship by an angle δ, and so the model responds to this extension history in the same way as the visco-elastic material shown in Fig. 7-3. We can derive the absolute modulus from Eqns. (7-14) and (7-15). These show the amplitude of the extension to be x_0 and that of the tension to be

$$x_0\left[\frac{E_1^2 + (E_1 + E_2)^2\omega^2\tau^2}{1 + \omega^2\tau^2}\right]^{1/2}$$

whence
$$|E| = \left[\frac{E_1^2 + (E_1 + E_2)^2\omega^2\tau^2}{1 + \omega^2\tau^2}\right]^{1/2}$$

It has now been shown that, in all three histories considered, the three-element model will exhibit the same qualitative behaviour as a visco-elastic material. However, this is not sufficient. We want to be able to choose a set of values of E_1, E_2, and τ so that the model will reproduce the behaviour of the given material quantitatively.

First, then, assume that a visco-elastic material does respond quantitatively in the same way as the above model. Its strain–time† relationship, when subjected to a constant stress, can be found by integrating Eqn. (7-10), giving

$$\varepsilon = \sigma\left(\frac{E_1 + E_2\{1 - \exp\left[-E_1 t/\tau(E_1 + E_2)\right]\}}{E_1(E_1 + E_2)}\right)$$

We eliminated the constant of integration using the condition that the stress was applied at zero time, when the strain instantaneously became $\sigma/(E_1 + E_2)$. The creep compliance, D_c, is therefore given by

$$D_c = \frac{1}{E_1} - \frac{E_2 \exp\left[-E_1 t/\tau(E_1 + E_2)\right]}{E_1(E_1 + E_2)} \tag{7-16}$$

where t is the time which has elapsed since the moment of application of stress. At very small values of t the value of the exponential term is unity and so

$$D_c = 1/(E_1 + E_2)$$

At very large values of t the value of the exponential term is zero, and so

$$D_c = 1/E_1$$

Hence, the variation of D_c with time is controlled by this exponential term;

† When dealing with the response of *models, force* and *extensions* were considered. Now that the equations are being applied to *materials, force* will be replaced by *stress* and *extension* by *strain*. The spring *stiffnesses* then become *moduli*, and the *viscous coefficients* of the dashpots become *viscosities*, in order to keep the equations dimensionally correct. The dimensions of the relaxation time are not altered. It becomes the ratio of a viscosity to a modulus, which is dimensionally the same as that of a viscous coefficient to a stiffness.

to show the variation of such a term with time, exp $(-x)$ is plotted against $\log_{10}(x)$ in Fig. 7-10. As this figure shows, most of the change in exp $(-x)$ with $\log_{10}(x)$ occurs in the range of $\log_{10}(x)$ from -0.75 to 0.25 [i.e., one unit of $\log_{10}$ (time)], or in the range of the value of x from 0.2 to 2.

In Eqn. (7-16), x has the value $E_1 t / \tau(E_1 + E_2)$, and so

$$\log_{10}(x) = \log_{10}[E_1 t / \tau(E_1 + E_2)] = \log_{10}(t) + \log_{10}[E_1 / \tau(E_1 + E_2)]$$

Thus, curves of

$$\exp[-E_1 t / \tau(E_1 + E_2)] \quad \text{against} \quad \log_{10}[E_1 t / \tau(E_1 + E_2)]$$

and of $\quad\quad \exp[-E_1 t / \tau(E_1 + E_2)] \quad \text{against} \quad \log_{10}(t)$

would be identical except that one would be displaced from the other along the $\log_{10}(t)$ axis. It therefore follows that if Eqn. (7-16) (derived for a three-element model) describes the behaviour of a visco-elastic material, then experimental values of D_c for such a material plotted against $\log_{10}(t)$ should give a curve in which most of the change in D_c occurs over one unit of $\log_{10}$ (time).

In Fig. 7-11, experimental values of D_c have been plotted against $\log_{10}(t)$, and it will be seen that the change in D_c occurs over a range of $\log_{10}$ (time) very much greater than unity. So it follows that the three-element model does not provide a quantitative description of the response of a visco-elastic material. This conclusion is emphasized by the curve of D_c against $\log_{10}[E_1 t / \tau(E_1 + E_2)]$ for a three-element model, plotted in Fig. 7-11.

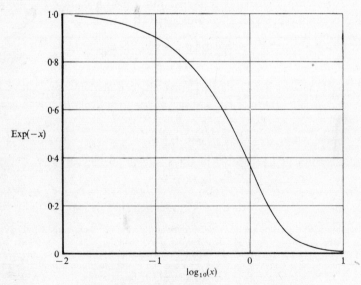

Fig. 7-10   Variation of exp $(-x)$ with $\log_{10}(x)$

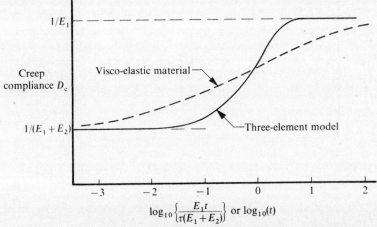

Fig. 7-11   Comparison of the three-element model with a visco-elastic material

Values of E_1 and E_2 were chosen so that the curves coincided at short and long times, and the curve for the model was displaced along the $\log_{10}(t)$ axis so that the curves for the model and the material crossed at the mid-point of the range of D_c (this effectively assigns a value to τ). A similar result would have been obtained had any other stress or strain history been investigated. We have to conclude, therefore, that a three-element model does not describe quantitatively the behaviour of a real visco-elastic material—the change in compliance of such a material extends over a much wider range of $\log_{10}$ (time) than that of the model.

7-3-4   The Wiechert model

Suppose now we construct a further model which has two Maxwell elements instead of one. Suppose, also, that the relaxation time of one of the elements is different from that of the other by a factor greater than 100. Since, from Fig. 7-11, most of the relaxation of a Maxwell element is completed in one unit of $\log_{10}$ (time), the relaxation of one of these elements will hardly have started before the relaxation of the other is almost complete. Thus, Fig. 7-11 can be regarded as illustrating the change in creep compliance due to the relaxation of the element with the shorter relaxation time. The entire curve of creep compliance against $\log_{10}$ (time) for this model will therefore be as shown in Fig. 7-12, where E_3 is the stiffness of the spring in the element with the shorter relaxation time. This figure shows that the change in creep compliance now occurs over a wider range of $\log_{10}$ (time).

If the difference in relaxation times of the two elements were less, a smoother curve changing over a narrower band of $\log_{10}$ (time) would have been obtained. Also, the width of this band could have been increased by

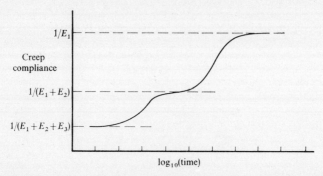

Fig. 7-12   Creep compliance of model containing two Maxwell elements in parallel

including still more elements. It is apparent, then, that the experimental curve of Fig. 7-11 could be reproduced by including a sufficient number of Maxwell elements with suitably chosen moduli and relaxation times.

This is the basis of the *Wiechert model* which has an infinite number of parallel elements, the relaxation times of adjacent elements differing infinitesimally. Hence, if $E(\tau)\,d\tau$ is the sum of the spring stiffnesses of those elements having relaxation times lying between τ and $\tau + d\tau$, the properties of the model can be represented by a graph of $E(\tau)$ against τ. This graph is the *distribution function of relaxation times*, or the *relaxation-time spectrum*.

In this model, we can regard the isolated spring as an element containing a dashpot of infinite viscosity, i.e., it has an infinite relaxation time. However, as it cannot be included in the distribution function without making this discontinuous at $\tau = \infty$, it is treated separately.

We could choose a distribution function of relaxation times so that a model having this distribution would reproduce the curve of creep compliance against time for the visco-elastic material shown in Fig. 7-11. We could then calculate the responses of this model to other stress and strain histories, and compare them with the experimental response of the material. If the method of describing visco-elastic properties in terms of relaxation-time spectra is satisfactory, the calculated and experimental responses should be the same, and so we now discuss methods of calculating the response to certain histories in terms of the relaxation-time spectrum.

7-4   Calculation of response from the relaxation-time spectrum
Consider first the case of stress relaxation. Using methods similar to those by which we derived Eqn. (7-9), the differential equation for the response of a single Maxwell element to any history is found to be

$$E_1\tau_1\frac{dx}{dt} = \tau_1\frac{df_1}{dt} + f_1 \qquad (7\text{-}17)$$

In a stress-relaxation experiment an extension is instantaneously applied, and kept constant with time. Since all the elements are in parallel, each one is held at the same extension, and so $dx/dt = 0$ for each one. Substituting in Eqn. (7-17), we get

$$\tau_1 \frac{df_1}{dt} = -f_1$$

which can be applied to any element in the model. Solving this equation gives

$$f_1 = E_1 x \exp\left(-t/\tau_1\right) \tag{7-18}$$

The constant of integration has been eliminated using the condition that the extension is applied at zero time and the force $E_1 x$ is developed instantaneously.

Now let Eqn. (7-18) be applied to those elements with relaxation times between τ and $\tau + d\tau$. Since these are in parallel, they are all at the same extension x, and the total force exerted by them all is the sum of the individual forces exerted by each one. Thus the total force, df, exerted by this group of elements is given by

$$df = xE(\tau) \exp\left(-t/\tau\right) d\tau$$

(Since all elements in this group have relaxation times between τ and $\tau + d\tau$, $\exp\left(-t/\tau\right)$ can be regarded as a constant for each. Thus the only quantity to be summed on the right-hand side of the equation is the stiffness of each spring in the group, and from the definition of the distribution function of relaxation times this sum is $E(\tau)\,d\tau$.)

We can now determine the total force, f, developed by adding together all the groups of elements of different relaxation times and the isolated spring of stiffness E_I, giving

$$f = xE_I + x \int_0^\infty E(\tau) \exp\left(-t/\tau\right) d\tau$$

The stress-relaxation modulus, $E_s = f/x$, is therefore given by

$$E_s = E_I + \int_0^\infty E(\tau) \exp\left(-t/\tau\right) d\tau \tag{7-19}$$

Thus, if we know the distribution function of relaxation times, we can determine the integral as a function of t, and find E_s.

Consider next the case of an extension varying sinusoidally with time, i.e.,

$$x = x_0 \sin \omega t$$

Since all elements are connected in parallel, at any instant they are all equally extended, and so this equation gives the extension in any element. If we consider first a single element, from Eqn. (7-17),

$$E_1\omega\tau_1 x_0 \cos \omega t = \tau_1 \frac{\mathrm{d}f}{\mathrm{d}t} + f_1$$

This differential equation can be solved to give

$$f_1 = \frac{E_1\omega\tau_1 x_0}{(1 + \omega^2\tau_1^2)^{1/2}} [\sin(\omega t + \delta)]$$

where $\qquad\qquad\qquad \tan \delta = 1/\omega\tau_1 \qquad\qquad\qquad$ (7-20)

Since δ varies from element to element, the tensions in the different elements will be out of phase with each other, and can only be added vectorially. However, the component of the tension *in phase* with the extension will be given by

$$f_1' = \frac{E_1\omega\tau_1 x_0}{(1 + \omega^2\tau_1^2)^{1/2}} \cos \delta \sin \omega t$$

and these components must be in phase for all elements, since the extensions of all elements are in phase. Substituting for $\cos \delta$ from Eqn. (7-20) gives

$$f_1' = \frac{E_1\omega^2\tau_1^2 x_0}{(1 + \omega^2\tau_1^2)} \sin \omega t$$

As in the case of stress relaxation, the tension in all elements can now be added to give the total tension in phase with the stress as

$$f' = (x_0 \sin \omega t)\left[E_\mathrm{I} + \int_0^\infty \frac{E(\tau)\omega^2\tau^2}{1 + \omega^2\tau^2} \mathrm{d}\tau \right]$$

where E_I is the stiffness of the isolated spring.

The in-phase modulus, E', is therefore given by

$$E' = E_\mathrm{I} + \int_0^\infty \frac{E(\tau)\omega^2\tau^2}{1 + \omega^2\tau^2} \mathrm{d}\tau \qquad\qquad (7\text{-}21a)$$

Similarly, the out-of-phase modulus, E'', can be shown to be

$$E'' = \int_0^\infty \frac{E(\tau)\omega\tau}{1 + \omega^2\tau^2} \mathrm{d}\tau \qquad\qquad (7\text{-}21b)$$

E_I does not enter into this expression since the tension in an isolated spring is in phase with the extension.

The determination of the creep modulus in terms of the relaxation-time spectrum is not so straightforward. The difficulty arises from the fact that, in creep, the total tension in the model remains constant, but the tension in an individual element can alter. Thus, $\mathrm{d}f_1/\mathrm{d}t$ in Eqn. (7-17) cannot be evaluated and so the equation cannot be solved. The difficulty can be overcome by postulating a different type of model. This consists of an infinite string of *Kelvin* elements in *series*. Each element is characterized by a spring stiffness and a *retardation time*, λ,† which is defined as the ratio of the viscous coefficient of the dashpot to the stiffness of the spring. It is found to be more convenient with this model to consider the contribution of individual elements to its *compliance*, rather than their contribution to its stiffness. If $D(\lambda)$ dλ is the sum of the compliances of the elements having retardation times between λ and $\lambda + d\lambda$, the graph of $D(\lambda)$ against λ is the *distribution function of retardation times*, or the *retardation-time spectrum*. There appears to be no generally accepted name for this model; let us call it the *series model*.

It can be shown that, provided such a model contains an element with zero retardation time (i.e., one with an isolated spring), then for any distribution of retardation times there exists an equivalent Wiechert model. Furthermore, the relaxation-time spectrum of this equivalent model can be determined in terms of the retardation-time spectrum. Conversely, it can be shown that, provided the Wiechert model contains an element of infinite relaxation time (i.e., an isolated spring), then for any distribution of relaxation times there exists an equivalent series model of which the retardation-time spectrum can be determined in terms of the relaxation-time spectrum. These proofs are, however, mathematically beyond the scope of this book.

From the above it follows that, if we can determine the creep compliance of the series model in terms of its retardation-time spectrum, then, since this can be transposed to the equivalent relaxation-time spectrum, we will have expressed the visco-elastic response to this history in terms of the same information as that used to describe the other histories.

The differential equation giving the response of a Kelvin model to any stress history can be easily derived:

$$Df = x + \lambda \, dx/dt \tag{7-22}$$

where D is the reciprocal of the spring stiffness. Since the tension in each of a string of elements in series must be the same, it follows that, if a constant tension is imposed on the model (as in a creep experiment), then the tension in each element must also be constant. Thus, if Eqn. (7-22) represents a single element in a model subjected to creep, it becomes

$$dx_1/dt = (D_1 f - x_1)/\lambda_1$$

and, since f is constant, this can be integrated to give

$$x_1 = D_1 f[1 - \exp(-t/\lambda_1)]$$

† The symbol λ is the one most generally used for retardation time. It has, however, already been used in this book for one of the Lamé constants. Confusion need not arise, since retardation times will only be mentioned in the present chapter, in which the Lamé constants do not appear at all.

The constant of integration was eliminated using the fact that the stress was applied at zero time, producing zero extension.

Consider now the elements having retardation times between λ and $\lambda + d\lambda$. Their total extension will be the sum of their separate extensions, the tension in each will be the same, and their retardation times can be considered the same. Thus, the contribution of these models to the total extension is given by

$$dx = fD(\lambda)[1 - \exp(-t/\lambda)]\,d\lambda$$

and the total extension by

$$x = f\left\{D_I + \int_0^\infty D(\lambda)[1 - \exp(-t/\lambda)]\,d\lambda\right\}$$

where D_I is the compliance of the isolated spring. The creep compliance is therefore

$$D_c = D_I + \int_0^\infty D(\lambda)[1 - \exp(-t/\lambda)]\,d\lambda \tag{7-23}$$

Equations (7-19), (7-21), and (7-23) give the stress-relaxation modulus, in-phase and out-of-phase moduli, and creep compliance in terms of the relaxation-time or retardation-time spectra. Thus, if these spectra are known,

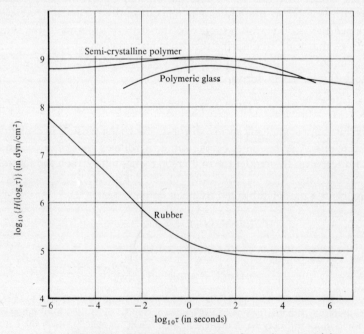

Fig. 7-13   Relaxation-time spectra for various materials

the responses can be calculated. Conversely, from the experimentally measured responses, the spectra can be determined, although the mathematics of the calculation are beyond the scope of this book. Hence, from each of the experimentally determined responses, the spectra can be calculated and compared. If they are identical we can say that this method provides a valid means of describing the visco-elastic properties of a material, since the one spectrum will then correctly predict the response of the material to any given history.

Materials have been studied in this way, and the results of these experiments show that the method is, in fact, valid. Examples of spectra are given in Fig. 7-13.

In this figure, the relaxation spectra have been plotted in a slightly different form from that defined earlier. The term $E(\tau)\,d\tau$ was defined as the contribution to the stiffness due to elements with relaxation times between τ and $\tau + d\tau$. An alternative is the quantity $H(\log_e \tau)\,d(\log_e \tau)$, which is defined as the contribution to the stiffness due to elements with relaxation times whose logarithms lie between $\log_e \tau$ and $\log_e \tau + d(\log_e \tau)$. This is often a more convenient form in which to express the spectra and can easily be related to the other form.

$H(\log_e \tau)\,d(\log_e \tau)$ is the area under a curve of $H(\log_e \tau)$ against $\log_e \tau$, between the ordinates $\log_e \tau$ and $\log_e \tau + d(\log_e \tau)$.

From the definition given, this must be equal to the area under a curve of $E(\tau)$ against τ between the same values of τ. Hence, since

$$d(\log_e \tau) = d\tau/\tau$$

when $\log_e \tau$ increases by $d(\log_e \tau)$, τ must increase in value by $\tau\,d(\log_e \tau)$, and therefore

$$H(\log_e \tau)\,d(\log_e \tau) = \tau E(\tau)\,d(\log_e \tau)$$

or
$$H(\log_e \tau) = \tau E(\tau)$$

Mechanical models have been used to develop the ideas presented in this and the preceding sections. However, it is important to realize that they are merely devices which have enabled us to derive Eqns. (7-19), (7-21), and (7-23). These equations, together with the relaxation-time or retardation-time spectra, describe the stress–strain–time response of a visco-elastic material. The equations and spectra are purely descriptions; they tell us nothing about the molecular origins of visco-elastic phenomena. The models have no significance other than that stated above; there is no analogy between the elements of material and the springs and dashpots of the models. It is, indeed, preferable now to think in terms of the equations and spectra rather than the models.

7-5 The superposition principle

Visco-elastic properties can be described by means other than relaxation and retardation-time spectra. An alternative approach is due to Boltzmann. Suppose at some instant of time, t_1, a stress $\Delta\sigma_1$ is instantaneously applied to a material. If stress and strain are linearly related, this stress causes a proportional instantaneous strain. This strain changes with time, so that at any subsequent instant of time, t, it is proportional to the stress $\Delta\sigma_1$ and dependent on some function of the elapsed time $t - t_1$, i.e., the strain $\Delta\varepsilon_1$ at time t due to a stress $\Delta\sigma_1$ applied at time t_1 is given by

$$\Delta\varepsilon_1 = \phi(t - t_1)\,\Delta\sigma_1$$

The term $\phi(t - t_1)$ is some function of the elapsed time which represents the 'memory' a material has at time t of some event which occurred at time t_1. It is therefore called a *memory function*. Suppose now that, at some later instant t_2, a further stress increment $\Delta\sigma_2$ is applied. Had the first stress not been applied, by the same argument as previously, the strain at time t would have been given by

$$\Delta\varepsilon_2 = \phi(t - t_2)\,\Delta\sigma_2$$

Now assume that *the increment in strain due to the second increment of stress is independent of the strain already existing in the material.* Further, let us assume that *the memory of a stress increment is unaffected by the application of any subsequent stress.* The strain at time t due to both stress increments is then given by

$$\Delta\varepsilon = \phi(t - t_1)\,\Delta\sigma_1 + \phi(t - t_2)\,\Delta\sigma_2$$

Since these assumptions imply that the response of a material to a complicated stress history is the sum of its responses to small parts of that history when each part is applied separately, they form the *principle of superposition*.

If a large number of stress increments is applied at different times, we get the equation

$$\varepsilon_t = \sum_{t_1=-\infty}^{t_1=t} \phi(t - t_1)\,\Delta\sigma(t_1)$$

where $\Delta\sigma(t_1)$ is the stress increment applied at time t_1. Suppose now that the material is subjected to a continuous stress history, and that between times t_1 and $t_1 + \Delta t_1$ the stress increases from σ to $\sigma + \Delta\sigma$. Then the equation

$$\Delta\sigma(t_1) = (d\sigma/dt)_{t_1}\Delta t_1$$

is obtained, where the suffix t_1 denotes that $(d\sigma/dt)$ is evaluated at time t_1. Substituting above, and letting Δt_1 become infinitesimal, gives

$$\varepsilon_t = \int_{-\infty}^{t} \phi(t - t_1)(d\sigma/dt)_{t_1}\,dt_1 \qquad (7\text{-}24)$$

This equation expresses the strain ε_t at any time t in terms of the stress history, $(d\sigma/dt)_{t_1}$, and a memory function, $\phi(t - t_1)$, which characterizes the visco-elastic properties of the material. As an example of its application, suppose a creep experiment is performed in which a stress σ_0 is instantaneously applied at $t = 0$. The stress history is then described by the equations

$$-\infty < t_1 < 0 \qquad \sigma = 0 \qquad d\sigma/dt = 0$$

$$0 < t_1 < t \qquad \sigma = \sigma_0 \qquad d\sigma/dt = 0$$

Hence, the integral in Eqn. (7-24) must be zero at all values of t_1 except the infinitesimal period at $t_1 = 0$, during which the stress is increased from 0 to σ_0. If this infinitesimal period is denoted by dt_1, substitution in Eqn. (7-24) gives

$$\varepsilon_t = \phi(t)\int_0^{dt_1} (d\sigma/dt)_{t_1}\, dt_1 = \sigma_0\phi(t)$$

Thus the creep compliance D_c is given by

$$D_c = \varepsilon_t/\sigma_0 = \phi(t) \qquad (7\text{-}25)$$

showing that the memory function, $\phi(t)$, can be obtained directly from a creep experiment.

Now suppose that at a later instant of time, $t = t_a$, the stress σ_0 is instantaneously removed. The stress history is then described by the equations

$$-\infty < t_1 < 0 \qquad \sigma = 0 \qquad d\sigma/dt = 0$$

$$0 < t_1 < t_a \qquad \sigma = \sigma_0 \qquad d\sigma/dt = 0$$

$$t_a < t_1 < t \qquad \sigma = 0 \qquad d\sigma/dt = 0$$

If these conditions are substituted in Eqn. (7-24), the integral will be zero at all times except $t_1 = 0$, and $t_1 = t_a$. As before, when $t_1 = 0$, its value is $\sigma_0\phi(t)$, and similarly, when $t_1 = t_a$, its value is $-\sigma_0\phi(t - t_a)$. Thus ε_t is given by

$$\varepsilon_t/\sigma_0 = \phi(t) - \phi(t - t_a) \qquad (7\text{-}26)$$

This equation is illustrated in Fig. 7-14. Had a creep experiment been performed, the curve OFADB would have been obtained. Suppose that, after performing such an experiment, we allow the material to recover, and repeat the experiment, except that we remove the load at time t_a. The curve OFAEC will then be obtained, and at time t, $\varepsilon_t/\sigma_0 = $ EG, and $\phi(t) = $ DG. Thus $\phi(t - t_a) = $ DE. However, from Eqn. (7-25), curve OFADB is a plot of $\phi(t)$ against time, and $\phi(t - t_a)$ is the ordinate of this curve at time $t - t_a$. Thus FH $= \phi(t - t_a)$ and so DE $= $ FH. It is therefore possible to predict the creep recovery from the creep curve. This prediction can then

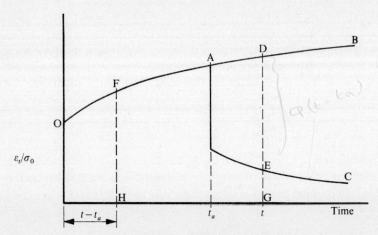

Fig. 7-14   Superposition principle

be tested experimentally, thus testing the principle of superposition experimentally. Many tests of this nature support the validity of this principle.

A series model will also conform to the principle of superposition. We have already derived the equations

$$x_1 = D_1 f[1 - \exp(-t/\lambda_1)]$$

and
$$dx/dt = (D_1 f - x_1)/\lambda_1$$

for a single Kelvin element, subjected to a constant tension f at zero time (Section 7-4). If at $t = t_a$ the tension, f, is removed, these equations become

$$x_{1a} = D_1 f[1 - \exp(-t_a/\lambda_1)] \tag{7-27a}$$

$$dx/dt = -x_1/\lambda_1 \qquad (t > t_a) \tag{7-27b}$$

where x_{1a} is the extension at time t_a. Solving Eqn. (7-27b), and using Eqn. (7-27a) to remove the constant of integration, gives

$$x_1 = D_1 f[\exp(t_a/\lambda_1) - 1] \exp(-t/\lambda_1)$$

which may be written

$$x_1/D_1 f = [1 - \exp(-t/\lambda_1)] - \{1 - \exp[-(t - t_a)/\lambda_1]\}$$

whence, for a series model,

$$x_1/f = \int_0^\infty D(\lambda)[1 - \exp(-t/\lambda)] \, d\lambda - \int_0^\infty D(\lambda)\{1 - \exp[-(t - t_a)/\lambda]\} \, d\lambda$$

This equation is of exactly the same form as Eqn. (7-26) and can be interpreted in the same way, using Fig. 7-14. It therefore follows that the two methods of representing visco-elastic behaviour are equivalent, and, by comparing Eqns. (7-25) and (7-23),

$$\phi(t) - D_{\mathrm{I}} = \int_0^{\infty} D(\lambda)[1 - \exp(-t/\lambda)]\, d\lambda$$

This enables us to determine the memory function from the retardation-time spectrum.

Instead of considering the strain in terms of the stress history, as we did to obtain Eqn. (7-24), we could consider the stress in terms of the strain history, giving

$$\sigma_t = \int_{-\infty}^{t} \psi(t - t_1)(d\varepsilon/dt)_{t_1}\, dt_1$$

where $\psi(t - t_1)$ is another memory function related to $\phi(t - t_1)$.

7-6 Assumption of linear behaviour

In representing the properties of visco-elastic materials with mechanical models, we used *linear* springs and dashpots. Similarly, when applying the principle of superposition, we assumed that a stress increment produced the same effect whatever the existing stress, which implies *linear* behaviour. Thus we have assumed that the stress–strain–time responses are *linear*, and consequently the theory which has been developed is called the *theory of linear visco-elasticity*.

The linear nature of the relationships are exemplified by the fact that the creep compliance and stress-relaxation moduli *at a given time* are constants independent of the stress or strain. Similarly, the in-phase and out-of-phase moduli *at a given frequency* are constants independent of the stress or strain. Had we calculated the stress as a function of strain when the material was extended at a constant rate, and drawn a curve, *each point of which was determined at the same time after commencement of extension*, we would have a straight line graph. (Such a curve could not, of course, be obtained by direct experiment, and is known as the *isochronal* stress–strain curve.)

This assumption of linearity in a visco-elastic material is equivalent to the assumption of Hooke's law in a perfectly elastic material. It is obeyed for real materials only at very small strains, and hence the theory developed in this chapter is applicable to real materials only at infinitesimal strain.

7-7 Worked example

The rubber whose relaxation-time spectrum is given in Fig. 7-13 is subjected to a stress-relaxation experiment. Determine its stress-relaxation modulus 1 s after applying the strain.

The stress-relaxation modulus is given by Eqn. (7-19) in terms of $E(\tau)$, whereas Fig. 7-13 gives the relaxation-time spectrum on a logarithmic scale. However, $H(\log_e \tau)\, d(\log_e \tau)$ is the contribution to the rigidity from elements with relaxation times whose logarithms lie in the range $\log_e \tau$ to $\log_e \tau +$ d($\log_e \tau$). Hence the equation

$$E_s = E_I + \int_{-\infty}^{+\infty} H(\log_e \tau) \exp(-t/\tau) \, d(\log_e \tau) \tag{7-28}$$

can be obtained, in the same way as Eqn. (7-19).

Figure 7-10 shows a curve of $\exp(-x)$ plotted against $\log_{10}(x)$, from which values of $\exp(-t/\tau)$ can be determined at different values of $\log_{10} \tau (t = 1)$. Multiplying each of these by $H(\log_e \tau)$ at the same value of $\log_{10} \tau$ (obtained from Fig. 7-13) gives the curve shown in Fig. 7-15. The integral

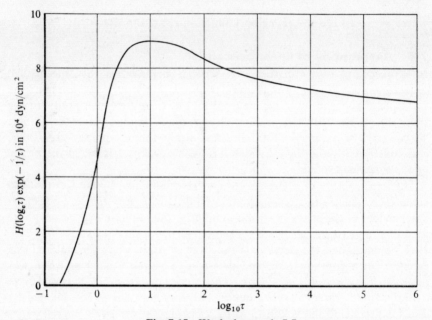

Fig. 7-15 Worked example 7-7

in Eqn. (7-28) is the area under this curve between the ordinates $-\infty$ and $+\infty$. It is clear from Fig. 7-15 that the major contribution to the modulus comes from relaxation times greater than the experimental time, and that (since for these relaxation times $\exp(-t/\tau) \approx 1$), unless $H(\log_e \tau)$ decreases sharply with increasing $\log_{10} \tau$ in this range, knowledge of the relaxation spectrum over very many decades of τ will be necessary. In fact, Fig. 7-15 shows that there is a considerable contribution to the integral from relaxation times greater than those given in Fig. 7-13, and so the problem cannot be solved. The value of E_I and whether or not it is significant compared with the value of the integral is also uncertain.

A similar difficulty would have been experienced had the calculation of E' at a value of $\omega \approx 1$ been attempted. However, since $\omega\tau/(1 + \omega^2\tau^2)$ approaches zero at high as well as low τ, E'' could be determined from knowledge of the relaxation spectrum over a much narrower range.

7-8 Summary

(i) If a stress is applied to a visco-elastic material there is an instantaneous strain. If the stress is held constant the strain increases with time. When the stress is removed, the strain instantaneously falls by an amount equal to the initial strain. The residual strain decreases with increasing time (Section 7-2).

(ii) If a strain is applied to a visco-elastic material, a stress is instantaneously developed. If the strain is held constant, the stress decreases with increasing time. When the strain is reduced to zero, the stress falls by an amount equal to that instantaneously developed, and so a negative stress is developed. This approaches zero with increasing time (Section 7-2).

(iii) If a stress which varies sinusoidally with time is applied to a visco-elastic material, the strain also varies sinusoidally with time, but is out of phase with the stress. The ratio of stress amplitude to strain amplitude, and the phase angle, both depend upon the frequency of oscillation of the applied stress (Section 7-2).

(iv) Qualitatively similar behaviour to that described in (i) to (iii) above is demonstrated by a model consisting of a spring and dashpot in series (a Maxwell element), this element being in parallel with a spring (Section 7-3-3). This model will not, however, quantitatively reproduce the behaviour of a real material (Section 7-3-3).

(v) To reproduce the behaviour of a material quantitatively, it is necessary to construct a model consisting of an infinite number of Maxwell elements in parallel (a Wiechert model), the relaxation time (the ratio of the viscous coefficient of the dashpot to the stiffness of the spring) of each element differing infinitesimally from its neighbours. If elements having relaxation times between τ and $\tau + \mathrm{d}\tau$ contribute $E(\tau)\,\mathrm{d}\tau$ to the total spring stiffness, the relation between $E(\tau)$ and τ is called the relaxation-time spectrum (Section 7-3-4).

(vi) If a relaxation-time spectrum is determined such that the Wiechert model responds quantitatively the same as a given visco-elastic material to one history, then it will respond quantitatively the same to any other history, and so the visco-elastic properties of a material may be described in terms of this model and the relaxation-time spectrum (Section 7-3-4).

(vii) It is difficult to calculate the response to some stress histories in terms of this model, and for these a model is considered which is made up of Kelvin elements in series. A Kelvin element comprises a spring and dashpot in parallel (Section 7-3-2), and is characterized by its retardation time (the ratio of viscous coefficient to stiffness). If elements having retardation times between λ and $\lambda + d\lambda$ contribute $D(\lambda)\,d\lambda$ to the total spring compliance of the model, the retardation-time spectrum is the relationship between $D(\lambda)$ and λ. For any retardation-time spectrum it is possible to determine the relaxation-time spectrum of a Wiechert model which would demonstrate identical visco-elastic behaviour (Section 7-4).

(viii) Models are only a means to a mathematical description of a material's properties. They have no relationship to the molecular mechanism responsible for the properties (Section 7-4).

(ix) The visco-elastic properties of a material may be described using a different, but equivalent, method due to Boltzmann (Section 7-5).

EXERCISES

7-1 A visco-elastic model consists of a Kelvin element having a spring of stiffness E_2' and retardation time λ in series with a spring of modulus E_1'. Show that it will have the same curve of creep compliance against time as the model illustrated in Fig. 7-8 if

$$E_1' = E_1 + E_2 \qquad E_2' = E_1(E_1 + E_2)/E_2 \qquad \lambda = \tau(E_1 + E_2)/E_1$$

7-2 A Wiechert model is extended at a constant rate of extension. Show that f, x, and t are related by the equation

$$f = \frac{x}{t} \int_0^\infty \tau E(\tau) \left[1 - \exp\left(-\frac{t}{\tau}\right) \right] d\tau$$

7-3 Show that the in-phase and out-of-phase moduli of a three-element model subjected to sinusoidally varying strain are given by

$$E' = E_1 + E_2\omega^2\tau^2/(1 + \omega^2\tau^2)$$
$$E'' = E_2\omega\tau/(1 + \omega^2\tau^2)$$

Show (a) that E'' is a maximum when $\omega = 1/\tau$; (b) that at this frequency the loss factor is given by

$$\tan\delta = E_2/(2E_1 + E_2)$$

and (c) that the slope of the curve of loss factor against ω is negative at this point.

8
Relationship between mechanical properties and molecular structure

8-1 Introduction

The preceding chapters have dealt with methods of describing relationships between the forces acting on solid bodies and the resulting deformations. A description of the main features of these relationships for typical materials has also been given. We saw that crystalline materials deform elastically at very small strains ($\sim$0·002), and plastically at greater strains. Inorganic glasses, which are non-crystalline, show similar elastic behaviour, but break before plastic deformation occurs. Rubbers deform elastically up to very large strains; while their bulk moduli are of a similar magnitude to those of crystalline materials, their shear moduli are smaller by a factor of about 10^{-4}. Polymeric glasses behave like inorganic glasses, and semicrystalline polymers like crystalline materials, except that all polymers are markedly more visco-elastic in shear than materials of low molecular weight. In this chapter we discuss the structure of these materials at a molecular level and relate their behaviour to this structure in a qualitative way.

8-2 Binding forces

Suppose two molecules in the gaseous state approach one another. Since, in an *ideal* gas, molecules interact only on collision, repelling each other, the forces between them will be as shown in curve 1 of Fig. 8-1(a). However, real gases differ in behaviour from the ideal, particularly at high compression. At very high compression the pressure rises above the ideal gas value. We would expect this because, under these conditions, the molecules are so close together that the repulsive forces shown in curve 1 of Fig. 8-1(a) are taking effect. At slightly lower compressions, however, the opposite occurs— the pressure falls below the ideal gas value. Hence, the forces shown in curve 2 of Fig. 8-1(a) must also be acting.

The net effect of both of these forces is shown in Fig. 8-1(b), from which we see that at a separation a_0 there is zero force between the two molecules. We can express the same idea more conveniently in terms of energy.

Suppose two molecules are separated by a distance a, where $a > a_0$. From Fig. 8-1(b), an attractive force exists between them and, to maintain the separation a, some external agency must hold them apart. If the molecules are now allowed to approach one another they must do work on this external agency; this work is derived from their own potential energy which will therefore decrease. Thus, the potential energy of the molecules is a function of their distance apart, and, from the above reasoning, we can calculate *changes* in this potential energy as separation changes; we cannot, however, express an *absolute value* for it. We avoid this difficulty by choosing an

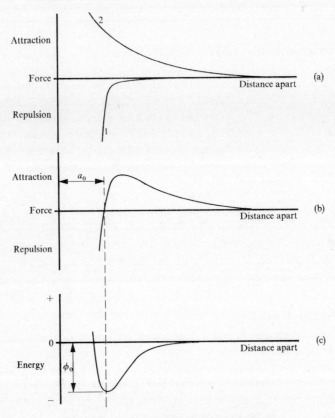

Fig. 8-1   Force of interaction between atoms

arbitrary separation and give the potential energy an arbitrary value at this separation. In fact, we take the potential energy between two molecules at infinite separation as zero. On this basis then, we define the *energy of interaction* of two molecules at a distance a apart [$\phi(a)$] as their potential energy at this separation. In Fig. 8-1(c), $\phi(a)$ is plotted against a.

From this figure, we see that at $a > a_0$, $\phi(a)$ is negative. This is because attractive forces operate at this separation, pulling the molecules together, and so *reducing* their potential energy (which we have said is zero at infinite separation). When $a < a_0$, repulsive forces act between the molecules and so work must be done on them to decrease this separation further, thus *increasing* their potential energy. Thus, $\phi(a)$ has a minimum value ϕ_0 at $a = a_0$, and since any system tends to adopt the state of least potential energy, we would expect the molecules to assume this equilibrium separation.

However, in the above discussion we have only considered the *potential* energy of the molecules, and neglected their *kinetic* energy. Suppose that when the separation is a_0 the molecules are moving apart. Their kinetic energy will be transformed to potential energy in overcoming the attractive forces between them, or, in other words, their speed of separation will decrease. If they are brought to rest before the force of attraction falls to zero, they will remain in association. If, however, their kinetic energy is great enough, they will still be moving apart when this force is zero and they will escape from each other's influence, i.e., they will dissociate. For this to occur, their kinetic energy at $a = a_0$ must be greater than ϕ_0.

The kinetic energy of the molecules of a gas decreases with decreasing temperature and, to be in a gaseous state, the molecules must be dissociated. Hence, if the gas is cooled, a temperature will be reached at which the mean kinetic energy of the molecules falls below ϕ_0. Molecules coming together will then remain in association at a separation a_0, i.e., the gas will condense. Conversely, when a solid is heated (i.e., the kinetic energy of its molecules is increased), it dissociates into a gas when the molecules have gained an energy of ϕ_0. The quantity ϕ_0 is therefore the *energy of dissociation*, or *binding energy*. It is given approximately by the latent heat of sublimation.

It is important to realize that all the molecules in a solid do not have the same, constant, energy. As in a gas, the energy of any given molecule fluctuates randomly with time, and there is a certain probability that it will have a given energy at a particular time. Thus, at any temperature, there is a probability that a molecule will dissociate from the solid, and it is this probability that increases with temperature.

It follows from the above argument that a gas should condense into a crystalline solid. If two molecules approach each other and remain in association, they take up the positions of lowest energy relative to each other. This pair will have potential minima at various positions around it, and other molecules which become associated with the pair will settle in these positions. Each time a molecule is added to the cluster it settles in a potential minimum and creates new potential minima into which other molecules can settle. In this way, a regular, repeating, crystalline pattern of molecules is built up.

If such a solid is deformed *dilatationally*, the distance between the molecules is increased. Thus, they must be moved to positions of higher energy and so work must be done (i.e., a force must act) on the material. When the force is released, the molecules move back to their equilibrium positions (i.e., the deformation will have been elastic). Since the time scale of these molecular movements is likely to be very small, visco-elastic effects are not likely to be appreciable, and so the deformation will be perfectly elastic. Since the work done in deforming the material is completely recoverable, it is equal to the stored elastic energy. This energy is therefore stored by increasing the energy of interaction (or, in thermodynamical terms, the internal energy) of the molecules.

If the solid is deformed *deviatorically*, the distance between the molecules remains the same, but neighbouring sheets slide relative to each other, as shown in Fig. 2-17. Thus, relative movement between molecules occurs at right angles to the line along which $\phi(a)$ is measured in Fig. 8-1(c). Since movement in this direction again raises a molecule from a potential minimum, this type of deformation will show similar elastic characteristics to dilatation. Now, the molecules lying in a sheet occur at regular positions in a pattern, and so the energy of interaction between this sheet and a molecule lying in an adjacent sheet will also show a regular pattern, i.e., the surface representing this energy resembles an egg-box, and molecules lie in the positions occupied by the eggs. Hence, if deviatoric deformation is allowed to continue until the molecules in a sheet pass through the potential energy maxima separating adjacent minima, they will settle in new equilibrium positions rather than return to their original ones. In other words, plastic deformation occurs.

Hence, by considering the deformation of a crystalline solid in terms of the energy of interaction between its molecules, we can explain the major experimentally observed features.

8-3   Relationship between binding energy and elastic moduli

From Fig. 8-1(c), when the force f acting on a crystalline solid is zero, the equilibrium spacing of its atoms is a_0. If f is not zero, but of such a value that the atoms are given a small† displacement u, their separation becoming $u + a_0$, then

$$f = \frac{d\phi(u)}{du}$$

where $\phi(u)$ is the energy of interaction at the displacement u. Since $\phi(u)$ is a continuous function, it can be expressed as the Taylor series

$$\phi(u) = \phi_0 + \left(\frac{d\phi}{du}\right)_0 u + \left(\frac{d^2\phi}{du^2}\right)_0 \frac{u^2}{2} + \text{higher terms} \qquad (8\text{-}1)$$

where ϕ_0 and all the differential coefficients are measured at $u = 0$. Since the energy of interaction is a minimum when the atoms are separated by the distance a_0 (at which $u = 0$), $\phi(u)$ is a minimum at $u = 0$, and so

$$\left(\frac{d\phi}{du}\right)_0 = 0$$

† The word 'small' is used to mean *small compared with a_0*, not small in absolute terms.

Also, for small displacements, third and higher order terms in u can be neglected compared with u^2. Thus Eqn. (8-1) becomes

$$\phi(u) = \phi_0 + \left(\frac{d^2\phi}{du^2}\right)_0 \frac{u^2}{2}$$

and

$$f = \frac{d\phi(u)}{du} = \left(\frac{d^2\phi}{du^2}\right)_0 u \qquad (8\text{-}2)$$

Since $(d^2\phi/du^2)_0$ is the rate of change of slope of the $\phi - u$ curve at $\phi = 0$, it is a constant, and so

$$f \propto u$$

which is Hooke's law. If Eqn. (8-2) had been expressed in terms of stresses and strains instead of forces and displacements, $(d^2\phi/du^2)_0$ would have been an elastic modulus. Thus the elastic modulus of a crystalline material is related to the sharpness of the minimum in the curve of energy of interaction against atomic separation.

To calculate the elastic moduli from Eqn. (8-2), we need to know the quantitative relationship between ϕ and u along the direction in which separation occurs during deformation. Also, since each atom is surrounded by other atoms (it does not have only one neighbour as visualized in the preceding discussion), we must know the energy of interaction with each of these nearest neighbours, and with more distant atoms within the range of its forces. The energies of interaction required for this calculation are known for very few crystalline materials, but when they are known, good agreement with experimental values is obtained.

8-4   Yield strength of crystals

It is possible to estimate the yield strength of a crystal in terms of its shear modulus, and we can do this most easily using a calculation due to Frenkel.

In Fig. 8-2(a), the line AB represents one plane of atoms, and the line CD represents an adjacent plane in the crystal. Suppose one atom is moved along the line CD in this plane, i.e., a shear deformation is applied to the crystal. In Fig. 8-2(b) a graph is plotted of the energy of interaction of this atom, $\phi(x)$, against its displacement, x, from its initial position, E. Suppose E is an equilibrium position of the atom, then the energy is a minimum at this point. From the shear modulus, we know the rate of change of slope of the curve of $\phi(x)$ against x at this minimum, so we can draw the portion HJ of the curve. When the atom is at the point F it is in an identical crystallographic position and so we can draw the portion of the curve KL, which must be identical with HJ. Similarly, the portion MN is identical with HJ and KL. The curve $\phi(x)$ against x must therefore be a periodically repeating function, with a period equal to b, the crystal spacing.

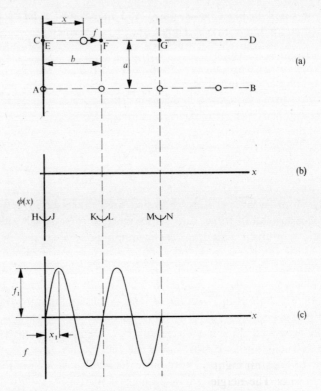

Fig. 8-2   Force acting on atom during shear deformation

We cannot know the nature of this function without detailed knowledge of the binding forces; we can only calculate the portions at the minima knowing the modulus. However, for a rough estimation, we can assume any periodic function applies; all such functions will give results of the same order of magnitude. Therefore, let us assume that

$$\phi(x) = -K \cos \frac{2\pi x}{b} \tag{8-3}$$

This gives $\phi(x)$ equal to $-K$ at $x = 0$, b, $2b$, etc., which meets the requirements of Fig. 8-2(b).

If f is the force which must be applied to the atom to hold it in equilibrium against the binding forces at a displacement x, then

$$f = \frac{d\phi(x)}{dx}$$

and, from Eqn. (8-3),

$$f = \frac{2\pi K}{b} \sin \frac{2\pi x}{b}$$

or
$$f = K_1 \sin \frac{2\pi x}{b} \qquad (8\text{-}4)$$

and so the force varies with displacement according to the curve shown in Fig. 8-2(c). From this curve, we see that at a displacement x_1 the force has a maximum value f_1, and that further displacement causes f to decrease, i.e., yielding occurs. Thus, the force required to cause a single atom to yield is this maximum f_1, which is equal to K_1. Therefore, if we can express K_1 in terms of the shear modulus, we will obtain an expression relating this modulus to the yield strength.

However, Eqn. (8-4) relates forces to displacements, whereas the yield strength is a stress, and the modulus relates stresses to strains. We must therefore change Eqn. (8-4) into a stress–strain relationship. From the definition given in Chapter 2 for small strain, the angle of shear, γ, is given by

$$\gamma = x/a$$

To change force to stress, we need first to determine the force required to displace every atom in the sheet CD instead of the single atom we have been considering so far. We obtain this force, f', by multiplying f by the number of atoms in the sheet. If we now divide f' by the area of the sheet, we get the shear stress σ acting on the crystal. These two operations affect only the value of K_1 in Eqn. (8-4); so we can rewrite this equation as

$$\sigma = K_2 \sin \frac{2\pi a \gamma}{b} \qquad (8\text{-}5)$$

where K_2 is now the yield stress of the crystal.

Equation (8-5) gives the curve of shear stress against angle of shear; the shear modulus, μ, is the slope of this curve at $\gamma = 0$. Since

$$\frac{d\sigma}{d\gamma} = \frac{2\pi a K_2}{b} \cos \frac{2\pi a \gamma}{b}$$

$$\mu = \frac{2\pi a K_2}{b}$$

and the yield stress, K_2, is given by

$$K_2 = \frac{\mu b}{2\pi a}$$

It is true that, because of the assumptions we made in arriving at this result, it is unlikely to be a correct numerical prediction of the yield stress. However, the error is unlikely to be more than one order of magnitude and so, since a and b are of the same order of magnitude, we would expect the yield stress to be smaller than the shear modulus by a factor of 10, approximately.

Experimental measurements of the shear modulus and the yield stress have been made for a large number of crystalline materials. For all of these, the yield stress is about 10^{-4} of the shear modulus, i.e., about one-thousandth of the value predicted above.

In the calculation we assumed that during elastic shear all the atoms were displaced simultaneously by an identical amount. Similarly, we assumed that, when yielding occurred, *all* the atoms *simultaneously* jumped into their new positions. If, instead of all the atoms moving simultaneously during

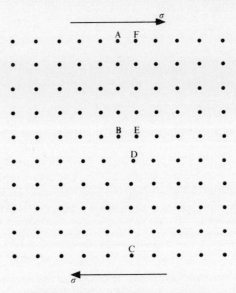

Fig. 8-3 Edge dislocation in a crystal

yielding, only a few moved at a time, this would greatly reduce the yield strength and could account for the large discrepancy between the predicted and observed values. However, if a crystal is perfect, i.e., every site is occupied by an atom, and none are anywhere else, this cannot occur. Before yielding can take place by the movement of a few atoms at a time, imperfections must exist in the crystal.

Consider, for example, the crystalline material illustrated in Fig. 8-3. (This diagram represents a cross section of the material in the plane of the paper; similar sheets of atoms exist above and below this plane.) The row of atoms AB (or plane in the three-dimensional block) terminates half-way down the crystal and can be regarded as an additional plane which has been inserted. This strains the lattice on either side, and around its termination B. Suppose now a shear stress is applied in the direction of the arrows. The atom at D jumps so that CDBA becomes a crystal plane, instead of CDEF, and EF becomes the inserted plane. (Since the crystal is three-dimensional, it is a row of atoms, not a single atom, which has moved.) Yielding thus occurs by the movement of a few atoms at a time, and the yield stress is many times smaller than that predicted for a perfect crystal.

The crystal fault illustrated in Fig. 8-3 is an *edge dislocation*. Other types of dislocation are also possible, and the plastic behaviour of crystals has been successfully interpreted in terms of the movement of these dislocations. This movement has also been observed directly by electron microscopy.

8-5   Structure and properties of inorganic glasses

From the discussion in Section 8-2, we would expect all substances condensing from the gaseous or liquid to the solid phase to crystallize. An inorganic glass is, however, amorphous, and so its failure to crystallize must be explained.

When a substance is crystallizing, atoms move relative to their neighbours until they are located on a crystal lattice site. The rate at which crystallization proceeds depends, therefore, upon the rate at which atoms find these sites. If they can move easily among their neighbours they will quickly find a suitable site and crystallization will proceed rapidly. If, on the other hand, their movement is hindered, crystallization will be an extremely slow process. In order for an atom to move relative to its neighbours, it must have sufficient thermal energy to overcome the binding forces, and at the same time there must be a suitable space for it to move into. In a glass, the probability of the simultaneous occurrence of these two events is small because firstly, the binding forces are large, and secondly, the molecules are large and irregularly shaped. Hence, the probability of a molecule finding a crystal lattice site is small.

As a material is cooled, its molecules lose thermal energy, and, because of thermal contraction, there is less space into which they can move. Hence, cooling reduces the probability of molecular movement, and temperatures can be reached at which this movement occurs infrequently during the time of an experiment. At these temperatures the material exhibits the mechanical properties of a solid rather than those of a liquid. If the rate of cooling is such that these temperatures are reached before the molecules have had time to find crystal lattice sites, then the substance solidifies in the amorphous state, i.e., it becomes a glass. One might expect a lower rate of cooling to lead to crystallization. This is true but cooling rates measured in fractions of a degree per century would be necessary to produce a crystalline solid from common glasses.

Clearly, then, the molecules in a glass, though not on crystal lattice sites, are held in position by binding forces in the same way as the molecules of a crystalline material. Deformation occurs, therefore, by the displacement of the molecules against these forces, and so the elastic properties of glasses are similar to those of crystalline materials. To exhibit plastic deformation, the molecules of a glass must completely overcome these binding forces and move to new sites. However, in glassy material the probability of such movement is very small, and while it is increased by the application of a force, this increase is sufficient to permit measurable plastic deformation only when the applied forces are large enough to fracture the material. Hence glasses fracture before yielding, i.e., they are brittle.

8-6   Origin of elastic forces in rubber

Rubber is an amorphous solid, and so might be expected to show similar elastic behaviour to glass. But, as we have seen, its behaviour is very different. Its shear modulus is smaller by a factor of about 10^4 (although its bulk modulus is of a similar magnitude), and, whereas a glass can be extended to a strain of only a small fraction of one per cent, a rubber can be extended elastically to several times its unstrained length.

If the deformation of a rubber in shear occurred by the displacement of atoms against binding forces, its shear modulus would be of a similar magnitude to that of crystals and glasses (as is its bulk modulus). Also, it would be unlikely to recover its original state from large deformation, since such deformation would move an atom through several interatomic spacings. The binding forces with its new neighbours would then be likely to make it settle in a new equilibrium position, causing permanent deformation. Clearly, then, the elastic forces resisting shear deformation of a rubber cannot arise from the displacement of atoms against binding forces.

We can show this by a further experiment. If a crystalline or glassy material is heated, its shear modulus decreases with increasing temperature. If, however, a rubber is heated, its shear modulus *increases* with increasing temperature. In fact, it is *directly proportional to the absolute temperature* (as is the pressure of an ideal gas). This suggests that the origin of elastic forces in a rubber is more akin to the origin of pressure in a gas than to the origin of elastic forces in crystals and glasses. As we will see shortly, this is, in fact, true.

Since elastic forces in a rubber do not arise because the atoms are displaced against binding forces, it follows that these atoms are likely to have a thermal energy greater than ϕ_0 [Fig. 8-1(c)]. This being so, we would expect them to move freely relative to their neighbours, and we would expect the material to have the properties of a liquid. Unvulcanized rubber, however, more closely resembles a solid; it can be made to flow as a liquid, but it is so extremely viscous that this property is not readily demonstrated. The high viscosity arises because the molecules consist of long chains of atoms. For unvulcanized rubber to flow as a liquid, entire chains must be displaced relative to neighbouring chains. This requires the co-operative movement of all the atoms in a chain and such movement is less likely to occur than the movement of inividual atoms.

Its behaviour as a very viscous liquid is changed by the process of vulcanization. In this process, molecules are 'tied' together chemically (cross-linked) at a few points along their length. The cross-links prevent the relative movement of entire molecules, but do not hinder the movement of short segments, so they change the material from a very viscous liquid into an elastic solid.

Thus, we visualize rubber as an assembly of long chain molecules, linked together at a few points along their lengths. Between these cross-links are lengths of chain, the atoms of which have sufficient thermal energy to move relative to their neighbours, provided, of course, that in so doing they do not disrupt the chain structure. We can now explain the elastic properties of rubber by reference to the model shown in Fig. 8-4. The posts A and B

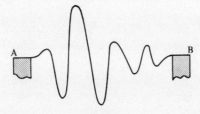

Fig. 8-4

represent cross-links, and the cord linking them represents the rubber molecule. The cord is in transverse oscillation (representing the movement of the molecule due to its thermal energy), which causes an inward force to act at the posts. If the energy of oscillation is increased (which is equivalent to increasing the temperature of the rubber), the inward force increases. The posts A and B can be moved a considerable distance apart against the inward force, i.e., large elastic deformations can occur, but these do not alter the energy of oscillation (or, in a rubber, the thermodynamic internal energy). If the posts are gradually moved farther apart, the force resisting further separation will increase greatly at the instant the cord is fully extended. This increase represents the difference between the shear modulus of the rubber and the value it would have if it were caused by binding forces.

The model can be made more realistic. Consider an isolated molecule whose shape, or *configuration*, is continually changing due to its thermal energy. As it assumes different configurations, so the length of the vector joining its ends, or its *end-to-end length*, varies. However, several different configurations will give the same value of end-to-end length, and a graph can be plotted of end-to-end length against the number of different configurations which will give this length. The form of this graph is shown in Fig. 8-5.

Suppose now, a segment of the molecule is held with its ends at a fixed distance *l* apart (as they would be at cross-links). Because of its thermal energy, the segment has a tendency, sometimes to assume a greater end-to-end length (i.e., to push the cross-links apart), and at other times to assume a shorter end-to-end length (i.e., to pull the cross-links together). From Fig. 8-5, it is clear that in this case the tendency to pull the cross-links together predominates (because the configurations with end-to-end length less than l are more numerous than those with end-to-end length greater than l), so there is a net tension between the cross-links. If the separation l was increased, the tendency to pull the cross-links together would be greater, and the net

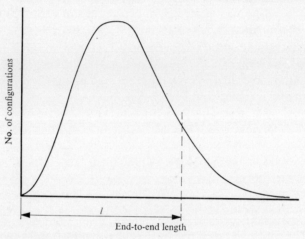

Fig. 8-5   Number of configurations of given length for long chain molecule

tension would increase. Conversely, if the separation was reduced, the net tension would decrease until it reached zero, at a value of l such that the tendencies towards tension and compression balanced.

As in the simplified model, increase in temperature would increase the tension at a given separation; the work done in deforming the material isothermally would not change its internal energy (again, as with an ideal gas); and provided the distance between the cross-links was short compared with the length of the molecule joining them, large elastic deformations would be possible.

The above discussion shows that the elastic forces resisting deformation of rubbers arise because of the thermal motion of the molecules. The pressure in a gas is also caused by the thermal motion of its molecules and so, clearly, both of these phenomena have the same cause. This explains the similarities between them that we have already noted.

Whereas all rubbers closely approach the model described, none conform to it exactly, and so a hypothetical material which does so is defined as an *ideal rubber*. It is possible to calculate the equation of the curve in Fig. 8-5; it has the same form as that for the distribution function of the velocities of the molecules of an ideal gas. From this equation, one can determine the force–extension relationship, and this is how Eqn. (6-1) was obtained.

In changing their configuration, the molecular chains are moving in a crowded environment. If the distance between cross-links is altered therefore (as during deformation), the molecules will take some time to assume configurations represented by Fig. 8-5. We would, therefore, expect noticeable visco-elastic behaviour.

8-7 Structure of glassy and semicrystalline polymers

We have seen that segments of molecular chains of a rubber have an energy greater than their binding energy. If, however, the temperature is reduced, a point is reached below which this is no longer true. The rubber must then either crystallize or form a glass. Which of these two it does depends upon its chemical nature, and upon the experimental conditions.

If natural rubber is cooled rapidly in the unstretched state, it forms a glass at about $-60°C$. The reasons for this are the same as the reasons for the formation of an inorganic glass—the atoms do not have time to find sites on the crystal lattice. When it forms a glass, the rubber changes dramatically from a highly extensible, low modulus, material to a brittle one with a shear modulus similar to that of an inorganic glass. This occurs over a temperature range of a few degrees. Also, as the temperature is further reduced, the modulus rises, whereas the modulus of a rubber decreases with decreasing temperature. These effects show that the configurational changes responsible for highly elastic deformation are prevented, and that, instead, deformation is opposed by binding forces.

If, on the other hand, the rubber is cooled slowly, it crystallizes. This is shown by its X-ray diffraction pattern, and by the fact that its density is higher than that of the rapidly cooled specimen. (Hence its atoms must lie more closely packed.) Both tests show, however, that the material is not completely crystalline.

The reason for this is that, whereas short segments of chain can move fairly rapidly and so find crystal sites, the larger lengths, which would have to rearrange themselves if the material crystallized completely, can only do so slowly, and so complete crystallization is prevented. The partially crystalline material can exist in one of two states; in one, the molecular chain elements outside crystals can change configuration easily (i.e., are rubber-like), and in the other, they cannot (i.e., are glass-like). However, the crystallization is the dominant feature controlling the mechanical properties, and there is less difference between the properties of these two states than between the properties of a rubber and a polymeric glass. The mechanical properties of a partially crystalline polymer are very different from those of a rubber—the shear modulus is much greater, and the material is ductile rather than highly elastic.

Whereas natural rubber can be cooled to form either a glassy or a crystalline polymer, other materials form only one type. Polymethyl methacrylate, for example, does not crystallize, whereas molten polyethylene cannot easily be cooled to form a glass. The reasons for this lie in the shapes of the atomic groupings along the molecular chain. In polymethyl methacrylate these groups are large and irregular, and so will not crystallize. At high enough temperatures both of these materials behave like unvulcanized rubber.

There is another way in which natural rubber can be made to crystallize. If it is deformed isothermally in simple elongation at room temperature, crystallization increases with extension. As extension proceeds, the molecules are pulled into straight and parallel configurations favouring crystallization. (This can also be regarded as an elevation of freezing point with stress— a one-dimensional Clausius–Clapeyron effect.) Thus, as extension of a rubber proceeds, its properties gradually change from those of a rubber to those of a crystalline material, and this is one reason why Eqn. (6-1) breaks down at large extensions. The material is no longer rubber-like at these extensions and so the stress required to increase its extension is very much increased. It also explains why a considerable amount of heat is evolved during extension; the formation of crystals releases their latent heat of fusion.

When a piece of rubber crystallizes under extension, the crystallization 'locks' the molecular chains in extended configurations. If the crystal melting temperature is above the temperature of the specimen, the chains remain in these configurations when the stress is released. In other words, the specimen is apparently permanently (or plastically) deformed. However, if the temperature is now raised above the crystal melting point, the chains are freed, and, provided no stress is applied, the specimen recovers its original dimensions. In a similar way, certain fluids penetrate into the molecular structure, 'loosening' it, and allowing the material to approach its original dimensions. This type of behaviour is demonstrated by any polymer crystallized in an extended state, and is a consequence of the long chain structure. This, then, explains the recovery from 'plastic' deformation of these materials, described in Chapter 6.

8-8 Summary

We have seen in this chapter that the elastic behaviour of crystals and glasses can be accounted for qualitatively by the binding forces between their atoms. To describe the plasticity of crystals we have to postulate the existence of defects, or dislocations, in the crystal lattice.

Elastic forces in rubbers have a different origin. These arise from the changes in molecular configuration which are continually taking place. Polymeric material can, however, also assume either glassy or partially crystalline states, and such changes modify its elastic properties dramatically.

The dislocation theory of crystal plasticity, and the theory of rubber elasticity, have both been developed quantitatively, although they have been expressed only qualitatively in this book. They both give good quantitative predictions of material properties. The exact molecular structure of glassy and partially crystalline polymers is, however, still a subject of some controversy, and so it is not yet possible to account quantitatively for the mechanical properties of these materials in terms of their structure.

Bibliography

Relationship between molecular structure and mechanical properties

COTTRELL, A. H. *The mechanical properties of matter*. John Wiley, New York, 1964.

FERRY, J. D. *Visco-elastic properties of polymers*. John Wiley, New York, 1961.

TRELOAR, L. R. G. *The physics of rubber elasticity*, 2nd edn. Clarendon Press, Oxford, 1958.

Mathematical theory of elasticity

GREEN, A. E. and J. E. ADKINS. *Large elastic deformation and non-linear continuum mechanics*. Clarendon Press, Oxford, 1960.

HEARMON, R. F. S. *An introduction to applied anisotropic elasticity*. Clarendon Press, Oxford, 1961.

LOVE, A. E. H. *A treatise on the mathematical theory of elasticity*, 4th edn. Cambridge University Press, London, 1927.

NYE, J. F. *The physical properties of crystals*. Clarendon Press, Oxford, 1956.

NOVOZHILOV, V. V. (Translated by J. K. LUSHER.) *Theory of elasticity*. Pergamon Press, Oxford, 1961.

Theory of elasticity applied to engineering structures

BENHAM, P. P. *Elementary mechanics of solids*. Pergamon Press, Oxford, 1965.

STIPPES, M., G. WEMPNER, M. STERN, and R. BECKETT. *An introduction to the mechanics of deformable bodies*. Prentice-Hall, New Jersey, 1961.

TIMOSHENKO, S., and GOODIER, J. N. *Theory of elasticity*, 2nd edn. McGraw-Hill, New York, 1951.

Answers to exercises

Chapter 2

2-1

	ε_{xx}	ε_{yy}	ε_{xy}	x' for point Q
	0·006	−0·0024	0·002	0·01
	0·002	−0·0013	−0·0025	−0·0075
	−0·0033	0·0025	−0·0022	−0·0088
	−0·0017	0·0025	0·002	0·012
	0·003	−0·002	0·0016	0·0016

2-2

		(a)	(b)	(c)
P	x	4·000		
	x'		2·985	
	y'		−0·012	0·015
	z'	0·006	0·006	
Q	y			
	x'	0·0225	−0·016	−0·009
	y'		4·012	
	z'	−0·010		
R	z			
	x'	0·012		
	y'		−0·018	−0·004
	z'	7·960		1·990
	ε_{xx}			0·001
	ε_{yy}	0·002		0·002
	ε_{zz}		−0·001	
	ε_{xy}	0·009		
	ε_{yz}	−0·004	−0·003	−0·002
	ε_{xz}		0·002	−0·004

2-3 $\varepsilon_{xx} = \varepsilon_{yy} = \varepsilon_{xy} = x/2a$

2-4 (a) $x' - x = 0·0305,$ $\quad y' - y = 0·0097$
    (b) (i) $y = 1·618x$
       (ii) $y' = 1·592x'$

2-5 Principal axes at $\theta = 76°43'$, principal strains 0·00347, −0·00547.

2-6 (a) Three, no two on the same straight line.

(b) The state of strain cannot be measured with such devices, irrespective of the number used and of the manner in which they are mounted.

2-7 47°52′,   1·0075 cm,   0·9975 cm.

2-9 (a) 12°38′

(b) $\varepsilon_{xy} = 4\cdot3 \times 10^{-4}$, $\varepsilon_x - \varepsilon_y = 10^{-3}$

(c) $\varepsilon_{xx} = -1$, $\varepsilon_{yy} = -11$, $\varepsilon_{xy} = -13\cdot3$, $\varepsilon_{1xy} = 2\cdot32$, $\varepsilon_z = -20$,

$\varepsilon_y = 8\cdot3$,   (all $\times 10^{-4}$). $\phi = 55°18′$.

2-10 (a) $\varepsilon_x = 98 \times 10^{-4}$, $\varepsilon_y = -8\cdot3 \times 10^{-4}$.

(b) 0·0107 radian.

2-11 (a) angle to positive x axis decreases by $18\cdot5 \times 10^{-4}$ radian; angle to negative y axis increases by $18\cdot5 \times 10^{-4}$ radian; angle to z axis unchanged.

(b) angle to x axis unchanged; angle to positive y axis increases by 5×10^{-4} radian; angle to positive z axis decreases by 5×10^{-4} radian.

(c) angle to negative x axis increases by 4×10^{-4} radian; angle to y axis unchanged; angle to positive z axis decreases by 4×10^{-4} radian.

2-12 (a) Angle between edge OZ and XY plane unaltered, angle between edges OX and OY increases by $2\cdot38 \times 10^{-3}$ radian;

(b) OX increases by $17\cdot5 \times 10^{-3}$ cm, OY increases by 9×10^{-3} cm, OZ increases by 75×10^{-3} cm,

(c) 7·2 cm³;

(d) $\varepsilon_1 = 3\cdot67$, $\varepsilon_2 = 4\cdot33$, $\varepsilon_3 = 0\cdot67$,   (all $\times 10^{-3}$).

2-13 Principal axes are diagonals of the square; principal strains are $\pm2\cdot5 \times 10^{-3}$.

2-14 (a) $-1, 0, +1$ (all $\times 10^{-3}$);

(b) 2×10^{-3} along each axis.

Chapter 3

3-1 $A = 20$ cm²; $\theta = 148°32′$;   normal stress $= 2\cdot96$ dyn/cm²;   shear stress $=$ 0·52, 0·80, 0·93 dyn/cm².

3-2 $\sigma_{xx} = -0\cdot034$; $\sigma_{zz} = 0\cdot979$; $\sigma_{xy} = 0\cdot171$; $\sigma_{xz} = -0\cdot0924$;

$\sigma_{yz} = 0\cdot183$; $F = 3\cdot94$; $x = 2$; $\alpha = 68°36′$; $\beta = 62°$;

$\gamma = 84°42′$; $\theta = 79°27′$; $\phi = 11°46′$.

3-3 Normal stress has maximum value of $F/\pi r^2$ when $\theta = 0$; shear stress has maximum value of $F/2\pi r^2$ when $\theta = 45°$.

3-4 Force is of magnitude $2\cdot56 \times 10^{11}$ dyn and acts parallel to the x axis in the positive direction.

3-5 Principal axes at 15° to coordinate axes by anticlockwise rotation; principal stress at 15° to x axis $= 7\cdot27 \times 10^8$ dyn/cm²; principal stress at 15° to y axis $= 10\cdot7 \times 10^8$ dyn/cm².

3-6 $F = 1.9 \times 10^{12}$ dyn; $\alpha = 70°40'$.

3-7

θ	ϕ	F (dyn)	
20°	329°41′	1·91	outwards
54°44′	324°44′	1·42	tangential
70°	126°6′	1·17	inwards

3-8 (a) $\sigma_d = 3$;   (b) $\sigma_{xd} = 1$ dyn/cm², $\sigma_{yd} = 3$ dyn/cm², $\sigma_{zd} = -4$ dyn/cm².

Chapter 4

4-1

OX	OY	OZ
3.9×10^{-3}	0.89×10^{-4}	-7.2×10^{-4}
1.22×10^{-4}	0.89×10^{-4}	0.56×10^{-4}
9.27×10^{-4}	13.3×10^{-4}	-12.5×10^{-4}
-37.0×10^{-4}	9.8×10^{-4}	12.4×10^{-4}

4-2 Angle increases by 5×10^{-4} radian.

4-3

	ε_x	ε_y	ε_z	σ_x	σ_y	σ_z	λ	μ	k
(a)				1·7	1·54	−1·24	0·8		
(b)	6·55	−4·86	−3·45						1·33
(c)					1·5	2·33	0·5	2·67

4-4

	ε_{xx}	ε_{yy}	ε_{zz}	ε_{xz}	ε_{yz}	σ_{xx}	σ_{yy}	σ_{zz}
(a)						−1·5	−4·5	−4
(b)	−4·30	0·69	0·97					
(c)		−2·3	1·6	−3·0	−1·0			

	σ_{xy}	σ_{xz}	σ_{yz}	λ	μ	k
(a)	2·5	−3·5	2·0			1·33
(b)	−1·35	1·80	−0·90		0·45	
(c)				0·273	0·5	0·606

4-5

	Change in length 10^{-3} cm			Change in angle 10^{-3} radian		
	OX	OY	OZ	OX and OY	OX and OZ	OY and OZ
(a)	3·0	−20	−3	−5	+7	−4
	−12·9	6·9	1·94	+3	−4	+2
	3·6	−23	3·2	−10	+6	+2
(b)	5·0	−8·0	−12·0	−5	+7	−4
	−21·5	2·76	7·76	+3	−4	+2
	6·0	−9·2	12·8	−10	+6	+2

Chapter 5

5-1 $\rho g l^2/2E$.

5-2 Axial stress $= g\rho x^3/3l^2$;   axial strain $= g\rho x^3/3El^2$.

5-3 Force applied at distance $r_1^2 E_1 d/(r_2^2 E_2 + r_1^2 E_1)$ from the bar of radius r_2.

5-4 (a) radius $> (W/2\pi^2 n\mu\varepsilon_1)^{1/3}$;   (b) radius $> (8W/15\pi^2\mu n\varepsilon_1)^{1/3}$.

5-5 $2Tl/\pi\{r_2^4\mu_2 - r_1^4(\mu_2 - \mu_1)\}$ radian;

inner portion, $U = lT^2 r_1^4\mu/\pi[\mu_2 r_2^4 - r_1^4(\mu_2 - \mu_1)]^2$ erg;
outer portion, $U = lT^2(r_2^4 - r_1^4)\mu/\pi[\mu_2 r_2^4 - r_1^4(\mu_2 - \mu_1)]^2$ erg.

5-6 $2l(T/\pi r^2 - fl)/\mu r^2$ radian.

5-7 C is at a height of $7\cdot5wl^4/EI$ cm above A, where w is the weight per unit length, and I cm^4 the second moment of area.

5-8 Ratio of stored elastic energies $(l/x)^3$, where l is the length of the beam.

5-9 Thickness $> (18Wh/EI\varepsilon_1^2)^{1/2}$ cm.

5-11 $\mu = 7\cdot94 \times 10^{11}$ dyn/cm^2;   maximum $\gamma = 7\cdot98 \times 10^{-4}$;   maximum $F = 6\cdot3 \times 10^6$ dyn;   maximum force under-estimated by about 2%.

Index